ISS: International Space Station

The Marvel of Human Ingenuity

D1607814

R. T. Kagels

Table of Contents

INTRODUCTION 1

CHAPTER 1: THE DREAM OF SPACE 5

THE ALLURE OF SPACE 6
Path to Reality 11

CHAPTER 2: THE MAKING OF A SPACE MARVEL 19

ENGINEERING FEATS 20
Global Collaboration 23

CHAPTER 3: LIFE ON THE ISS 29

LIVING IN ZERO GRAVITY 30
Working in Space 37
Cultural Exchange 42

CHAPTER 4: SCIENTIFIC BREAKTHROUGHS 45

SPACE MEDICINE 46
Earth and Environmental Studies 51
Astronomical Research 53

CHAPTER 5: INSPIRING THE NEXT GENERATION 55

EDUCATIONAL OUTREACH 56
Media and Public Perception 59
The Role Models in Orbit 62

CHAPTER 6: TECHNOLOGICAL MARVELS ON THE ISS 65

ADVANCED SYSTEMS AND MACHINERY 66
Sustainability in Space 68
Communication and Navigation 71

CHAPTER 7: THE FUTURE OF SPACE STATIONS 75

BEYOND THE ISS 76
New Frontiers 78
Challenges and Opportunities 81

CHAPTER 8: RISKS AND CHALLENGES OF LIVING IN SPACE..................**87**

SPACE DANGERS 88
  Emergency Protocols 93

CHAPTER 9: THE UNSEEN ISS **97**

BEHIND THE SCENES 98
  Cultural and Personal Life 100
  The Art of Space Living 103

**CHAPTER 10: THE FUTURE OF THE ISS—MISSIONS AND NEW
DEVELOPMENTS** **107**

EXCITING NEW MISSIONS 108
  Goodbye to the ISS 110

CONCLUSION **113**

BONUS CHAPTER: STRANGE FACTS ABOUT THE ISS **115**

AUTHOR BIOGRAPHY **117**

REFERENCES **119**

IMAGE REFERENCES 136

Introduction

We live in a divided world that is rife with conflict. From politics to religion, climate change to human rights, there is no one topic everyone can agree on. At times like these, symbols become particularly important, not to mention meaningful. Symbols are unique to them; they can instill hope among people of differing opinions and help unite them in myriad ways, even in the most turbulent of times. Only a few symbols worldwide are capable of such a feat, of course. Undoubtedly, the International Space Station (ISS) is the chief among them. How could that not be the case, given its history?

In many ways, the ISS was born from an era that's just as divisive and turbulent as the one we live through. The years before the ISS was founded were colored by tension and conflict. World War II had come to an end, leaving the world battered and broken in many ways. Nations and their people had picked themselves back up and started rebuilding, only for tensions to rise again. Those tensions had translated into the Cold War, a period of fear where war, it seemed, could erupt anew at any given moment between the United States and Soviet Russia. Yet, what arose out of that period was not war, though there were some close calls. Instead, it was the space race.

The space race was a literal race between the US and the USSR to leave Earth and go to space. The two countries launched their respective space agencies to meet this goal. Quickly after that, they turned their goal into something a bit more concrete: reaching the moon. The moon essentially became a destination for mankind to reach, one that was immensely hard to get to, to put it mildly. The US and USSR aimed to be the first to touch the moon, and their race would be dogged. There would be times when the USSR achieved various firsts before the US ever did. However, the US would soon catch up and soon surpass them. In due time, US astronauts would get into orbit and land on the moon. Neil Armstrong and Buzz Aldrin would thus become the first men on the moon.

You would think that losing the space race to the US in this fashion would end the USSR's space exploration ambitions. It did not. It made them settle on a new goal: building their own space station and placing it in orbit. Naturally, this lofty goal was also picked up by the US, and both countries succeeded in achieving it. By this point, the tensions between them had cooled significantly, to the point that nuclear war was less an imminent danger and more an avoided catastrophe. This meant the relationships between the two countries and their respective space programs. Their relationships would improve so much that the two agencies, NASA and Roscosmos, would start working together.

Seeing as the space race and NASA itself were the products of rising tensions and hostilities between two rival nations, no one could have expected this outcome. Similarly, no one could have expected NASA and Roscosmos to enter into an official agreement following the Columbia disaster, which cost many NASA astronauts their lives and resulted in NASA shutting down their shuttle program. This is when Roscosmos agreed to carry NASA astronauts up to space with their spacecraft. Where in space, you ask? Why, to the ISS, of course.

Around this time, NASA and Roscosmos decided that they should merge their two space stations and thus create the ISS. The news, when announced, took the world by firestorm, and how could it not? The fact that two nations who had been at each other's throats could now collaborate in this way was nothing short of staggering and inspiring. The news was so inspiring that other space agencies of other nations also decided to get in on the action. Over time, the ISS became the definition of "international," with some 30 countries becoming members.

With all this being the case, you can see how and why the ISS is considered such a huge symbol of hope these days in light of the various conflicts around the world. The fact that Russian cosmonauts and American astronauts continue to collaborate on the ISS amidst the tensions over the Russia-Ukraine war is an immense symbol of hope. The ISS, however, is not just a symbol of hope. It is a beacon of human ingenuity. A technological marvel, it is home to some of the most staggering scientific discoveries made in the past few decades. Not only that, but it is also a pioneer that has opened up the way to the new age of space exploration, which will be defined and colored by not just one but many space stations. To understand this new era, though, we must first understand the ISS, how it came to be, and how it led the way. Only then will we understand all that will come in the next new age of space exploration.

Chapter 1:

The Dream of Space

Space is to place as eternity is to time. -Joseph Joubert

The first person ever to reach space, the Soviet cosmonaut Yuri Gagarin, did so in 1961. Thus, it has been about 63 years since that momentous occasion (*A Brief History of Space Exploration*, n.d.).

Space exploration is a relatively new concept when you think about it. After all, it has not been a century since humans first touched the stars. The first person ever to reach space, the Soviet cosmonaut Yuri Gagarin, did so in 1961. Thus, it has been about 63 years since that momentous occasion (*A Brief History of Space Exploration*, n.d.). Just because space exploration is new to us humans, though, does not mean that we have only just developed an interest in space itself. On the contrary, space has been a source of fascination for us since the dawn of humanity. The only reason we waited so long to start exploring space is that it took us a while to develop the technology to do so.

The Allure of Space

This might be a bold claim to make from specific perspectives. If humankind was interested in space, they could have begun their exploration efforts far earlier. However, such people only have to look at the history of astronomy and even astrology—given that it is rooted in ancient civilization's attempts to understand celestial bodies like stars, the moon, and the sun—to grasp this. You see, humankind became interested in the sky and, therefore, in space thousands of years ago, before the first telescope was conceived.

Truthfully, we need more information on when ancient civilizations first started trying to decipher the movement and meaning of celestial beings like the sun. However, one of the earliest examples of an ancient civilization doing so was found in Northern Europe. This example, from an 11.8-inch bronze disc depicting the sun, stars, and crescent moon, is from 1,600 BC. Known as the Nebra disk, it serves as clear evidence that mankind has long been trying to understand the universe, even when they didn't yet know what the word "universe" meant.

It must be noted, of course, that ancient civilizations' attempts to understand the universe were more religious than scientific. Such civilizations usually believed that the sky or the heavens were the domains of gods and goddesses. They therefore controlled the stars, sun, and moon. For instance, anything unusual these civilizations observed in the sky, like a solar eclipse, was a sign sent to them by the gods. In the 21st century, we know this is not the case. Yet some of those old beliefs remain, which is why astrology is a thing, prompting people to say things like "Jupiter is in retrograde," whatever that means.

Some archeologists believe that the Nebra disc isn't a religious artifact but rather an early attempt at a calendar. It may also have been a rudimentary attempt to understand the planet's movements. If so, it's remarkably similar to the petroglyphs, that is to say, cave drawings that various Native American tribes of yore have left behind, depicting the astronomical phenomena that they have observed throughout the years.

It took civilizations a while to keep written records of such phenomena, but they did start doing so in due course. The earliest known example of this is from Babylonia circa 1600 BC. Their records keep careful track of the changing positions of planets, phenomena like eclipses, and more. Similar records have been found among ancient Chinese, North European, and Central American cultures and civilizations. Given that, scientists often consider astronomy to be the first science.

It took some time for this science to start properly developing, though, just as it would take mankind some time to reach space once they began their efforts in earnest. The first real developments in astronomy occurred in ancient Greece, around 500 BC. This is when Greek philosopher Thales developed a basic framework to predict when eclipses would happen. Meanwhile, Plato completed the first-ever calculations to determine the rough sizes of the Earth, moon, and sun. Eratosthenes and Aristarchus used mathematical and geometric formulas to figure out their rough distances from each other around 250 BC. Eratosthenes also calculated the Earth's circumference. In time, thanks to these calculations, he would come to be known as the father of geometry.

Other astronomers quickly took the fore. Hipparchus created the first-ever star catalog and started naming the constellations he identified. We use many of these names to this day. In the year 330 BC, Heraclides proposed the first solar system model, which only included five planets since that was all that could be seen to the naked eye without a telescope. This model suggested that the Earth rested at the center of the solar system instead of the sun. Other solar system models came after, some placing the sun at the center, but this erroneously persisted for many years.

Then, in 200 AD, Ptolemy took the stage and suggested that the planets move in irregular, circular motions. He went on to write a 13-book series covering astronomical phenomena. These books were then used by Hipparchus, who created the first star map. Many other discoveries were likely made following these. Unfortunately, most of them were lost in the fire in the Library of Alexandria in 272 AD. Many of the works held in the library were singular in that there weren't any other copies of them since the age of the printing press had yet to come. So, sadly, we have no way of knowing just how much knowledge was lost to us in this incident. We do not know why astronomical studies stopped for a couple of centuries after that, however, hey would pick back up again in the Renaissance.

One of the people who revitalized astronomy during the Renaissance, specifically in the 1500s, was Copernicus. Born in 1520, Copernicus actively challenged the notion that the Earth was at the center of the solar system. This was a dangerous claim to make at the time because it was what the church supported, and, back then, people who went against the church could find themselves at an inquisition trial, ending in either their execution or imprisonment. This was a fact that Copernicus would discover later on, but that is a story for another time.

In any case, Copernicus reintroduced the idea that the sun was positioned in the center of the solar system. Many other astronomers, including Tycho Brahe, who built the Danish

Observatory in the 1580s, picked up his theory. His student, Kepler, formulated the Laws of Planetary Motions based on what he learned and observed firsthand at the observatory in the 1600s. Thus, he figured out that planets moved in elliptical circles around the sun and even devised a mathematical formula to calculate their orbits.

Kepler's formula became the mathematical definition of the solar system we live in until Galileo picked it up. Using them, Galileo developed the laws of motion. He also built the first-ever telescope and was able to observe the sun, moon, and the Milky Way in a way that no one had ever done before. With that telescope, he discovered that Jupiter had its moons and Venus had phases, to name a few of his accomplishments.

Galileo's discoveries fed into Newton's later on. He used what he had learned from them when developing the law of gravity and the laws of accelerated motion. While at it, he created the first reflecting telescope, invented calculus, much to the chagrin of many future high school students, and devised the theory of light.

All this may seem inconsequential compared to feats like orbiting the Earth, flying to the moon, and living on a space station. However, those accomplishments could not have been made without them. Still, these accomplishments were decidedly confined to the Earth, and that situation would not change until the 20th century. Specifically, it wouldn't happen until after World War II, when new technologies, like rockets, would make attempting such feats possible.

Funnily enough, mankind's first attempts to reach space would be born out of another threat of war, one between the United States and Soviet Russia. Following World War II, the two nations would grow increasingly uneasy with one another and eventually enter into the Cold War, where they would both prepare for possible nuclear war with one another. Such a war being a possibility, they would strive to develop their rocket and satellite technologies to attack one another and defend themselves to the best of their abilities. The ensuing space race— a literal race to see who could reach space and then the moon faster—would directly result from these attempts.

The space race culminated in the US win, though it looked like they were lagging behind for a time. The first nation to get an unmanned spacecraft into orbit was NASA. They did so by launching Sputnik into space in 1957. Thus, Sputnik became the first artificial satellite ever to orbit the Earth. A month later, Soviet Russia launched Laika, the (unfortunate) dog, into space. Thus, Laika became the first space dog. Meanwhile, the US was struggling to catch up. They attempted to launch their artificial satellite into orbit, but their efforts did not succeed until 1958, a full year after Russia's Roscosmos.

This year, after the Explorer satellite was successfully launched, NASA, the US Federal Space Agency, was formed. Thus, the US kicked up its space exploration efforts up a notch. The Soviets, however, still seemed to be in the lead. In 1961, they became the first country to get a living, breathing man—Yuri Gagarin—into space. Three weeks later, the US and NASA achieved the same win, a sign that they were closing the gap between themselves and Russia. The first American man to ever get to space was Alan Shepard. Following Shepard's achievement, President John F. Kennedy set a new endpoint or goal for the space race: the moon.

Kennedy essentially made a speech tasking NASA with getting mankind to the moon. NASA took to the challenge with gusto. Yet, NASA still had some catching up to do. In 1959, Russia's Roscosmos, not NASA, got the first unmanned spacecraft, the Luna 2, to the moon. It is similar to Roscosmos' Valentina Tereshkova, who became the first woman in space and the first person to perform a spacewalk there.

These achievements might have been disheartening for NASA, but they proved to be anything but. Every achievement that Roscosmos made only stoked the fire burning within NASA, so they committed themselves to the Apollo Program, which was all about getting man to the moon. Without going into too much detail, NASA first reached the moon without landing in 1968 and 1972, respectively. It did so with two lunar modules that orbited the moon with astronauts on board. Then, in 1969, Neil Armstrong and Buzz Aldrin became the first men to set foot on the moon, thus officially winning the space race for the United States. Obviously, the race to the moon wasn't as straightforward as it sounds here. If anything, it was very turbulent, filled with anticipation, a rush to beat the clock, and even tragedy. That, however, is not the subject of this book. If you want to learn more about the Apollo Program and the space race, you can check out another book I've written on the subject matter: the *Kennedy Space Center.*

Path to Reality

If you think winning the space race ended space exploration, you must think again. Reaching the moon may have signified the end of the space race, but at the same time, it marked the real beginnings of space exploration. Following this achievement, NASA and other space agencies worldwide doubled their efforts to make discoveries. To that end, NASA launched various probes into space in the 1960s and 70s. These probes, dubbed Mariner, gathered data on planets like Mars, Venus, and Mercury.

While NASA focused on collecting these kinds of data, Russia's Roscosmos had gone down another path entirely: Building the first-ever space station. The first space station in the world, the Salyut, was built by Soviet Russia and put into orbit in 1971. Originally, it was only meant to stay in orbit for six months. The first crew to board the Salyut was comprised of the Russian cosmonauts Georgi T. Dobrovolski, Vladislav N. Volkov, and Viktor I. Patsayev. Their crew stayed on board for 24 days, a record (Uri, 2021).

Following the launch of the Salyut, Roscosmos made many different iterations of the space station throughout the ensuing decades. Each model was more advanced than the next, and this fashion continued until the Mir station was put into orbit. The Mir Station is the most elaborate space station ever to be put into orbit. It also has the distinction of being the longest-lasting space station ever. The station consists of four modules, and its first module, Kvant 1, was put into orbit in 1987. The Kvant 2 followed two years later, in 1989. Then came its penultimate module, Spektr, in 1995 and its last module, Priroda, in 1996 (*Mir FAQs - Facts and History*, 2001).

Just because the space race was over didn't mean that the US and Russia were drawn inspiration from one another. Since Russia was so invested in space stations, it was natural that the US would also become interested in them. Inevitably, the NSA and the US decided to create their space station, too. This station became known as Skylab. As America's first space station, Skylab launched into orbit in 1974. The launch proved far from idyllic, though. Precisely 63 seconds after Skylab launched, its micrometeor shield opened up. This shield was originally supposed to protect the station from space debris and act as a thermal blanket. That it was compromised was problematic. After all, without it in place, the heat within the station would rise and slow-cook any astronauts on board.

Luckily, NASA could get the station temperatures to drop by adjusting its attitude in orbit. While NASA scientists worked on this solution, which took them a while to come up with, the launch of the crew initially boarding the station was delayed by 10 days. Upon arriving at the station, the crew had several problems to fix, like the communications array and the fact that the solar array had been ripped off during the launch. To hear NASA officially tell it, they were so frustrated that they forgot that communications had been restored with ground control and ended up expressing their frustration by using four-letter words—the ones that begin with "F" and "S," in case it wasn't obvious—and ground control had to remind them that communications were back online multiple times (Howell, 2018).

Despite these hitches, Skylab was successfully put into orbit and has hosted many crews for years. Thus, the US, like Russia, was able to realize a dream that was centuries old when you think about it. The first space stations might not have been launched until the 1970s, but the idea of them has been around since the 1800s, thanks to a little genre called science fiction or sci-fi. A space station was first featured in a sci-fi novel in 1869. The book in question was Edward Everett Hale's *The Brick Moon*. In the book, a satellite made of brick spheres and arches—hence the name—is kept aloft thanks to a massive, moving wheel, similar to a hamster wheel.

This design wasn't very realistic and was never even attempted. The first book to have a realistic depiction of what a space station might look like was Konstantin Tsiolkovsky's *Out of the Earth*, published in 1920 (*Space Stations*, 2022). The space station in this station was replete with a greenhouse and artificial gravity, which was generated by having the station spin on its axis. This effect was replicated in later, real-life space stations.

The next fictional iteration of the space station came in 1949. It appeared in a book called *The Conquest of Space* by William Ley. Next came Robert Heinlein's *Space Cadet* in 1948 and Arthur C. Clarke's *Islands in the Sky* in 1952. In the 50s, the space station was treated as a pitstop or way point between longer destinations. Many sci-fi books were written at this time, all featuring stations in Earth's orbit and that of the moon and other planets. Real life has yet to fully catch up to sci-fi in this regard when you think about it. Space stations exist in the Earth's orbit, but one has yet to be placed into that of the moon or, say, Mars. Yet doesn't equal "never", though. There are numerous plans to establish a space station in the moon's orbit and other space stations to come.

Sci-fi wasn't the only genre of literature to feature space stations before they were built. Other genres, like non-fiction, tackled the concept as well. The first non-fiction book to do so was Herman Potocnic's *The Problem of Space Travel*. Published in 1929, this book actively tried to work out the mechanics of a space station and how it might function. It also introduced the idea of having the station rotate on its axis using a rotating wheel. In essence, it got the rotation part right but not the wheel bit, seeing as the space stations of today—yes, there are more than one—are sadly devoid of big hamster wheels.

As you can see from all this, people were thinking about space stations and how they might be built long before building them. This is part of why space agencies worldwide focused on developing the technologies and approaches needed to translate space stations into real-life entities. Take docking, for instance. Scientists needed to develop the necessary technology for docking, and astronauts needed to perform the movement in space. Otherwise, they would never be able to board and live in the space stations that would be created (Chao, 2013).

To that end, the US' first official docking mission took place in 1966. The person leading the mission was Neil Armstrong. The mission was a success, thus making all future docking missions possible. This was just the first manned docking mission, though. Just a year later, scientists proved that unmanned docking missions were possible, too. The ones to prove this feat were not NASA, though. Instead, it was Russia's Roscosmos. Thanks to their operation, the first unmanned docking mission in space occurred in 1967 and ended in resounding success.

Technological advancement was essential for making space stations a reality. This advancement, however, didn't end once the first station was created. Instead, scientists always looked to improve their stations, so the Salyut had many different iterations before being replaced with Mir. Now, not all Salyut models were drastically different from one another. Salyut 1-5, for instance, was pretty much the same, at 65 feet by 13 feet. Salyut 6 and 7, on the other hand, were newer models compared to the previous five iterations of the space station. Their designs were so different that these two models were considered the second generation of the Salyut. One key difference was that Salyut 6 and 7 had two docking ports. These ports allowed the station to be refueled and resupplied by unmanned cargo ships.

Unlike the Salyut stations, Mir had a modular design, allowing six additional modules to be added to the station over the ensuing ten years. The station may have held up for a remarkably long time, but it started experiencing many problems as time passed and it aged. At one point, a major fire broke out on the Mir, and while the crew managed to get it under control, it served as a clear signal that it was perhaps time to retire Mir. By this point, in 2001, plans were already underway to launch the International Space Station. The US and Russia had been collaborating for some time, with Russia shuttling American astronauts to space and even hosting several of them on Mir. At long last, Mir was de-orbited and eventually crashed into the Southern Part of the Pacific Ocean.

Fun Fact: The Soviet Union dissolved while cosmonaut Sergei Krikalev was on board the station. Given this fact, Krikalev has become known as the Last Soviet Citizen.

The United States and Russia aren't the only countries to have built their very own space station. China has as well. China's space station is Tiangong, which means Heavenly Palace. Unlike Mir or the ISS, for that matter, Tiangong 1 had a single-module design. Launched in 2011, it measured 34 by 11 feet and was considered by China to be a prototype. Unlike this prototype, the current iteration of the Tiangong is a three-module model. These modules that make up the station were launched in 2021 and 2022 (Howell, 2018b).

The Tiangong may be an impressive feat by any technological standard, but it pales slightly compared to the ISS. This is partly because the ISS is much bigger than the Tiangong, consisting of 17 modules. This makes an abundance of sense, given that five space agencies from 15 countries are a part of the ISS. The ISS is the largest man-made space structure and has cost $100 billion to build. Building this massive structure naturally took years. The first module making up the structure was built in 1998 by Russia. This was the Zarya module. The various space agencies involved in the station's construction launched numerous other modules after its launch. Thus, the ISS grew to be the size of a football field over the years.

The ISS became operational long before it was fully "completed." On November 2nd, 2000, astronaut Bill Shepherd and cosmonauts Yuri Gidzenko and Sergei Krikalev traveled to the station. Thus, this international crew became the first individuals to set foot aboard the ISS. They spent four months on board working to "activate" the ISS properly. This marked the beginning of perpetual human presence in ISS and space. This collaboration also really brought the symbolism inherent in the ISS to the fore.

Meanwhile, construction and additions went on. In 2001, for instance, a US lab named Destiny was attached to the station, thereby increasing its living space by 41%. To this day, Destiny remains the main research hub of the ISS. In 2008, the European Space Agency (ESA) added the Columbus Lab to the station. A month later, the Japan Space Agency added its module, Kibo, to the ISS (Chao, 2013).

Thus, the ISS grew with each addition and contribution, becoming a true hub of interstellar innovation. Since 2012, 125 different spacecraft have been launched to travel to the ISS for many reasons. How exactly did this happen, though? How did all these countries and agencies come together to forge this international center of development and hope? How did they even construct it, and how does the ISS work? Let's find out.

Chapter 2:

The Making of a Space Marvel

The International Space Station is to work on peaceful projects. In space, we're all people from Earth. —Sunita Williams

Space exploration may have started as a race between two nations, one driven by political strife and even potential war. However, the ISS was a unifying force between nations and a symbol of peace. No symbol is ever made in a single day, though. Rather, it takes time for one to form, especially one that is powerful and meaningful. This goes double for an engineering marvel such as the ISS. There's no question that the ISS is an incredible achievement, technologically speaking. Its size alone is staggering to think about. The station orbits the Earth at an altitude of 250 miles. That's not that far when you think about it.

At the very least, it's close enough for the ISS to have a Houston area code, in case you want to call it (Achenbach, 2013).

At this height, it can still feel the effects of the atmosphere of the Earth, which causes it to experience some drag. The ISS is boosted regularly to prevent this and to keep the station from veering off course. This takes a massive amount of energy, which the station can generate. That's not the only marvel it's capable of achieving. The station also self-cools using ammonia. This is potentially dangerous since ammonia is lethal enough to kill a human being with a single inhale, and a leak could spell catastrophe for the crew members on board the ISS. Then there's pressurization. The ISS could experience rapid depressurization should debris puncture through it, causing the crew members on board to die very sudden deaths. The station, however, has measures to prevent such issues while maintaining its day-to-day operations and keeping itself afloat in orbit, following the trajectory it's meant to follow.

Engineering Feats

To understand how the ISS can do all this and more, we have to first grasp how it's structured and made. To start, the ISS is made of 16 different modules, as you know, but it can be divided into two distinct parts. The first part of the ISS comprises its habitable modules, which are the ones where the crew members on board live, sleep, eat, and work. The second part of the ISS houses the technical necessities of the station,—these are the components necessary for the station to maintain its temperatures, have the power to keep functioning, and the like. Hence, this is where the solar arrays powering the station and the radiators lie.

The habitable part of the station is made up of modules that were each individually sent up to orbit. Once in position, they docked on the ISS, thus becoming a part of it. Despite being added onto the station later on, like puzzle pieces being fit in place, these habitable modules are sealed airtight. They have to be if human beings are to survive in them. Otherwise, they would let all the air and warmth within them out, proving it non-habitable.

The habitable part of the ISS's volume is 13,696 cubic feet, not including the spacecraft that visit and dock the station (Howell, 2018a). It has seven sleeping quarters, though more can be added periodically, particularly when incoming crews take over various duties from outgoing ones. The habitable part of the ISS also has a gym, a very important part of the station, as you'll learn later on, two bathrooms, and a cupola, which provides crewmembers with a 360-degree view of the Earth.

When added together, the habitable and non-habitable parts of the ISS weigh 925,335 lbs. in total. Again, this does not include any docked spacecraft on the station. The non-habitable part of the station consists of its solar arrays, heating and cooling units, and radiators. Before the solar arrays were installed, ensuring these components ran without hitches was challenging, particularly on space stations that predated the ISS. Solar arrays were first installed on Russia's Zarya space station. This was in 1988, making Zarya the first station to possess this technology. Later, Zarya would become the ISS's first module, but more on that later.

Returning to the solar arrays, these panels are the ISS' main power source. The ISS rotates on its axis as it travels around the Earth. So, you might be thinking that there are times when the arrays don't get direct sunlight. This, however, isn't the case. You see, had this been the case, the station could not have been powered as well or efficiently as it is. The engineers working on these arrays knew this, so they made it so that the solar arrays on the station rotated as well. The ISS panels are always turned toward the sun, no matter where the station is in the Earth's orbit.

Currently, the ISS isn't powered exclusively by the same solar arrays it started with. Those are being supplemented by a newer, more advanced model installed in 2021. This model, known as Roll-Out Solar Arrays (ROSA), features panels that are much more compact yet more powerful than standard solar arrays. These arrays roll out the way a carpet would, and they produce 30 kilowatts of power per single panel (Patel, 2022).

The ROSAs were installed on the ISS before the old solar panels were fueling them. The two sets of arrays generate a combined power of 250 kilowatts. That they can do so has prompted NASA to commit to using ROSAs in future projects, including the Lunar Gateway. Once that station is finished, its Power and Propulsion Element (PPE) will be fueled entirely by two ROSAs.

The old solar panels on the ISS were mostly pre-assembled and then launched into space. Some, though not many, were assembled by astronauts in orbit. The newer ones were rolled out and added to the station by astronauts in a seven-hour-long mission. This wasn't the only time astronauts had to go on a mission to take care of solar arrays. Sometimes, the batteries of solar arrays need replacing. This is a task taken care of by astronauts on spacewalks. These are often long and challenging missions because of the lack of gravity involved, and the bulky spacesuits astronauts have to wear make things cumbersome.

The ISS radiators are officially known as the Active Thermal Control System (ATCS). Many systems on the ISS produce excess heat (*Active Thermal Control System (ATCS) Overview*, 2022). This heat must be controlled so the crew members and equipment on board don't cook. To that end, the ATCS does three things: heat collection, transportation, and rejection. The system collects all the excess heat, transports it to the circulating ammonia loops outside the structure, and cools it down. This is a delicate operation for two reasons. The first is that space is an exceedingly cold environment, to put it very mildly. Therefore, maintaining a certain temperature in the space station is of the utmost importance. The second is that ammonia is a dangerous substance. Should there be a leak because the space station got punctured by something, for instance, inhaling the ammonia that slips into the air inside the ISS can rapidly kill whoever happens to be there. Luckily, the ISS has systems on board to prevent this potential risk. One of these systems is purely for performing evasive maneuvers that can get the ISS out of the way of incoming meteors or space debris, as you will discover later on.

There is one other component of the heating-cooling system on the ISS, and that's its reflective panels. The ISS has certain reflective panels on it. Their job is to direct the heat they receive from the sun, which the station is pretty close to when you think about it, away from it. This prevents the ISS from overheating. While this is happening, the station's cooling systems circulate the air inside, thus making sure things always remain at a comfortable temperature (Wagenen, 2020).

Global Collaboration

All that's well and good, but how was the ISS built? The short answer to that question is "bit by bit." How about the long one? The ISS was assembled in orbit, module by module, thanks to an array of spacecraft, astronauts, and cosmonauts willing to spacewalk to join the necessary modules and the wonders of robotics. The modules were linked to one another using the connection nodes found on them. Modules would be assembled like so:

1. A module would be launched up into space using rockets.

2. Two modules would be put into position to align their connecting nodes.

3. Those nodes would be linked together.

The Zarya was the first module launched into orbit to start the construction of the ISS. Russia's Zarya was sent to space on November 20th, 1998, using a proton rocket. The US launched the Unity module into orbit two weeks later using the STS-88 shuttle. The astronaut on board the shuttle performed several spacewalks to link the two modules together. Today, the Zarya module houses the station's fuel and supplies its battery power.

The third module to be added to the ISS was Zvevda. This service module was attached in 2000, and like the first module, it was created by Roscosmos and launched into space using a proton rocket. The next addition to the station came in 2001 when the Destiny Laboratory Module was attached to it. This was followed up with the Leonardo Multi-Purpose Logistics Module and the Raffaello MPLM. As you can see from those names, adding these modules marked the "artsy" phase of the ISS' construction.

The various modules that came to make up the ISS are all different sizes and shapes, which means that the ISS lacks the uniformity that a suburban neighborhood, with homogenous homes lining the block, would. Some of its modules are triangular. Others are more canister-shaped. Still others are beam-like, spherical, or resemble small, flat panes. Out of all these shapes, astronauts and cosmonauts live in spherical or canister-shaped units. Meanwhile, the ISS' backbone, called the truss, is a beam-like structure with numerous triangular shapes sticking out of it in a row. There are also panel-shaped components to be found here; these are usually the solar panels that power the station.

As the years went by, more modules kept being added to the station. 2008, for instance, the Japan Space Agency contributed the first pressurized component of its Kibo Laboratory. The second component of the lab came just months later, along with a useful robotic arm. The last module to be added to the station is the Bigelow Expandable Activity Module. This was attached to the ISS in 2016 and was taken up into space by a SpaceX Dragon.

Aside from the one that Japan installed, the station's main robotic arm is a component known as Canadarm2. Canadarm is more formally known as the Remote Manipulator System, though no one bothers with that name. Everyone opts for the first one, especially since Canada installed it. The installation process of Canadarm took quite a bit since it's made up of 18 different components. When fully extended, Canadarm is 55.7 feet long. The part of the arm meant to grab things is called the end effector and not "the claw." Sadly, there won't ever be tiny green Martians exclaiming, "Oooh, the claw," upon seeing it. This is probably because the end effector looks less like a claw and more like a cylindrical hand. An arms control unit aboard the ISS controls it and Canadarm in general.

The ISS was assembled through 36 separate shuttle flights and six Russian Soyuz or proton rocket launches. When construction was finished, at long last, in 2016, the ISS weighed as much as 330 cars. It was also 243 feet by 361 feet, which means it's as big as a football field with its end zones. Meanwhile, the surface area of its solar panels, including the newer ones, is 27,000 square feet. Currently, the ISS is the third brightest object in our solar system, the first two being the sun and the moon. At its current orbit speed, it passes over the same spots on Earth once every three days but can circle the Earth in 90 minutes, meaning it orbits the Earth 16 times in 24 hours. That, in turn, means that the astronauts on board see 16 sunsets and sunrises every day.

While the ISS is both a massive engineering feat and a beacon of hope for many, its future seems a little uncertain of late, especially in light of the current goings on down here on Earth. You probably already know that Russia's war with Ukraine is still, as of the writing of this book, ongoing. You also probably know this has caused tension between Russia, the United States, and its Western allies. You might not know that this tension can be felt all the way up in the ISS and might impact its future, seeing as Russia is contemplating pulling out of the ISS after 2024. Should they decide on this action, they will inevitably build their national space station instead.

Russia first threatened to pull out of the ISS after the Ukraine War in response to the sanctions Western countries placed on it during this time. The threat is not yet taken as fact since Russia has yet to notify NASA that it will officially be pulling out. Why 2024 and not sooner? This is because the ISS was originally only supposed to last until 2024 (Wall, 2023). The plan was to be de-orbited after that year, but the US and its various allies on the ISS have been unwilling to do so. They've argued that the ISS has a couple more years left in it yet, and it does, at least from an engineering-technological standpoint. As for how long those "few years" amount to, the US government is considering keeping the ISS in orbit until 2030, assuming their international partners agree. Russia may not. They may do exactly as they've threatened and pull out. This would be problematic as the ISS is designed so that its different components are reliant on one another to continue functioning. If Russia pulls out, then the station will inevitably have to be reconfigured as it will lose parts.

Scientists and engineers will find themselves scrambling to figure out how to keep the station in good working order, with it missing some part or another.

That Russia might pull out from the ISS is understandable, to a degree, given the political situation at hand. However, this is not as easy a decision as some might think. After all, pulling out of the ISS means giving up the myriad of benefits it has to offer its member countries (*Benefits for Everyone from International Space Station Research*, 2022). One of these benefits, for instance, is that the ISS allows scientists from different parts of the world to collaborate in a way they never could have managed down on Earth. This allows for collaborative research, studies, and investigations to be launched by different parties working together, leading to new finds, developments, inventions, and innovations much more quickly. Another benefit is that the ISS improves peer-reviewed scientific research methods and the results they obtain while reducing the cost and expense of the studies and investigations on the station. The ISS can do this by dividing those costs and expenses among the different groups participating in them rather than having one nation or institution shoulder it all.

There are specific scientific benefits to the ISS, as well. One is that scientists on the station can share specimens and create open data repositories. Another is that they avoid unnecessary duplication of experiments. The ISS even benefits scientists who've never been to the station by providing them with new data obtained through experiments conducted on the station. Students down on Earth have access to all this data, too, and even participate in some studies in collaboration with the ISS and work with students and scientists worldwide without having to move away from where they live or travel to space. Add to that how the ISS maintains multilateral collections of research results and makes all this information publicly available to everyone on nasa.gov. It's easy to see how scientifically beneficial the ISS is for people across the globe.

Naturally, the ISS doesn't just offer scientific benefits to people. It has had myriad historical and political benefits, too. The ISS first emerged as the brainchild of President Clinton and then-president of Russia Boris Yeltsin at the Vancouver Summit in 1993 (Oberhaus, 2020). Following the summit, the US invited Russia to become a full partner in the ISS project. One of the historical benefits of Russia joining the ISS was that it incentivized Russia to abide by the Nuclear Non-proliferation Act by presenting them access to many technological advancement ambitions while giving their scientists a golden opportunity to work on exciting projects unrelated to the military. In this regard, Russia's involvement in the ISS can be considered a good ending and true resolution to the Cold War. Because of this, the ISS can also be considered a first example of how space exploration can be seen and used as a diplomatic tool.

Chapter 3:

Life on the ISS

It was a strange lightness, a drifting feeling. Zero gravity. I understood that everything that once seemed solid and immovable might just float away. — Lisa Unger

The ISS might not precisely be a touristic destination, but it is a place perpetually habited by human beings. There's always some person or another staying at the station. It must noted, though, that each crew only stays on board for a couple of months before switching out with another crew, meaning that the ISS isn't anyone's permanent home. That may sound a little odd to some. After all, why wouldn't the crew stay for longer periods, say a year, if the ISS is such a great place to be? The short answer to that question is that it's not safe. To clarify, I don't mean that the ISS is an inherently unsafe place to be. Space, however, is owing to several factors like radiation. Now, the ISS takes many measures to ensure that the crew members staying on board are safe during their stays. One of those measures is limiting how long they are allowed to remain on the station. Such measures ensure that crew members remain healthy during their time in space. At the same time, though, they add certain quirks to space-living.

Living in Zero Gravity

The thing about living and working in space is that it means working in microgravity environments because gravity doesn't exist in space. Microgravity is an exceedingly weak form of gravity, the kind that's only found in orbiting spacecraft like a space station. Gravity is the force with which a planet, satellite, or some other such object draws the things that are on it or near enough to it toward its core or center. How much gravity an object, say a planet, exerts on an object, say the human body, depends on its mass and the distance between it and the human body. The bigger a planet is, the more mass it has. The more mass a planet has, the more gravity it can exert on the human body. Take the moon, for instance. The moon is many times smaller than the Earth. As such, its gravity is much weaker than Earth's, which is why the moonwalk—the Neil Armstrong kind, not the Michael Jackson kind—is a thing (Munson, 2023). Alternatively, take Jupiter. Jupiter is many times bigger than the Earth. As such, Jupiter's gravity is many times more powerful than Earth's, enough to crush whatever spacecraft flew into its atmosphere (*Jupiter: Facts*, 2023).

Gravity also has to do with distance. The farther you get from an object with a gravitational pull, like the Earth, the weaker the effect of gravitational pull on you. Now, the ISS is in Earth's orbit, as you know. The reason it's able to remain there is partly because it's kept there by Earth's gravitational pull. However, the

ISS is, technically speaking, in space. It's distant enough from the Earth that the effects of gravity for anyone on board the station are much, much lighter than they would be on Earth. This is why there's microgravity on the ISS. So, what does microgravity feel like?

According to the astronauts and cosmonauts at the station, living in microgravity feels a bit like floating. The things around them appear to float as well. If an astronaut were to leave a pen in the air, for example, the pen would stay airborne and slowly start floating away. The reason the pen would float away has to do with the fact that the ISS is orbiting the Earth at a speed of 4.5 miles per second (Cranford & Turner, 2021). This applies a certain degree of force to the center of the objects and people inside, causing them to drift. Coupled with the weak gravitation pull of the Earth on the station, this makes crew members feel like they're in perpetual freefall when they're on the ISS. You know what they sort of feel like if you've ever been on a roller coaster or experienced sudden turbulence on an airplane. So, you can imagine that it can be a challenging feeling to get used to.

Crew members living on the ISS then feel weightless during their time there. Knowing this, they often train for hours and hours on end to get used to the feeling so that they can live and work properly while at the station. US astronauts typically train for this with NASA's reduced gravity flight program. The reduced gravity flight program is not exactly a pleasant experience. The astronauts who've participated in it have dubbed it the "Vomit Comet," which should tell you as much (Nola Taylor Redd, 2017). They've chosen to give the program a pleasant name because the maneuvers that the aircraft they're in perform to get them to experience zero gravity have caused a fair share of them to puke up their guts. You see, the Vomit Comet flies at 30 to 40 parabolas, which is ridiculously fast for those not in the know. It climbs high into the sky, and as it does, astronauts feel the effects of gravity 30 to 40 times more powerfully than they do on Earth. Once the Vomit Comet reaches its summit, it plunges. As it does, astronauts get to experience the feeling of weightlessness for 25 seconds. Then, the plane starts climbing again to repeat the entire process.

> **FUN FACT:** The longest time an astronaut has spent feeling weightless/in a zero-reduced gravity environment was 438 days. The astronaut in question was Valeri Polyakov.

As unpleasant as this program may sound, astronauts must train in it and get used to the weightlessness they'll experience on the ISS. Otherwise, they wouldn't be able to function properly on board. Even a seemingly simple task such as eating dinner would become challenging for them. Imagine that you're an astronaut and want to drink some water. You pull out a package of water— not a bottle because all liquid consumables come in packages to reduce their risk of floating away upon opening. You open your package, and bubbles of water immediately start floating away. If you were an astronaut, you would have been trained to catch those water bubbles midair and drink, thus preventing dehydration. You would have received similar training for crumbly foods that are liable to float away similarly. Likewise, you would have been trained to handle liquid chemicals, which you would have to experiment with on the ISS. If you hadn't received this training, you wouldn't have been able to conduct the experiments you were meant to conduct on the ISS, at least not safely.

These aren't skills astronauts can figure out while on the Vomit Comet. Luckily, astronauts undergo another training program at NASA to learn such things: neutral buoyancy training. Neutral buoyancy training takes place in the Neutral Buoyancy Lab (NBL). The NBL is a massive swimming pool, holding 6.2 million gallons of water (*NBL Facilities*, 2017). Station in the Weightless Environment Training Facility (WETF), which can be found at the Johnson Space Center at NASA, the pool is 25 feet deep. Once at the facility, astronauts don adapted space suits with various weight and floatation devices that create a neutral buoyancy. Neutral buoyancy feels pretty close to the weightlessness they'll experience in microgravity. So, astronauts will dive into the pool with life-sized mock-ups of the devices and systems they must use at the ISS. Using these, they'll practice the various moves they'll have to execute and experiments they'll have to run. As they do so, the devices that they are using will float around them since they're in the water, much as they will in space. Thus, astronauts will get used to using floating equipment, tools, and devices before they ever set foot on the ISS.

Radiation is another factor that astronauts must get used to in space. However, there are stars, like the sun, and their light travels massive distances. Light has a fair bit of radioactivity. It's known, officially, as cosmic radiation. We are exposed to this radioactivity back on Earth but are protected from it by the Earth's magnetic field. Space, however, has no such magnetic field to protect against cosmic radiation, which is why the crew members on the ISS are regularly exposed to 30 times the radiation levels they'd normally be exposed to on Earth. For reference, that is the equivalent of a year's worth of radiation that those of us who are Earth-bound would get.

Radiation is a huge risk to us human beings. It can cause all sorts of problems, like radiation sickness and a wide variety of cancers. Given that, NASA and its member space agencies have taken many measures to protect their astronauts and cosmonauts from space radiation. To that end, they've ensured that the ISS is made with certain protective materials that lower the radiation the astronauts on board are exposed to. The keyword is "lower," as no material we know of can completely negate radiation, meaning that crew members are still exposed to some of it on the station. Once they step out of the station to perform spacewalks—something they occasionally have to do for maintenance operations, like when they added new solar panels to the ISS—they are exposed to more of it (*Radiation and Life*, n.d.).

Now, cosmic radiation levels aren't consistent in every part of space. Different parts of the ISS are made with different materials with varying protective shielding levels. Some parts of the ISS have cement shielding, which stops ionizing radiation. The most used parts of the ISS, like the crew quarters and the galley, where they eat and hang out, are shielded with polyethylene, which is good for stopping a broad spectrum of radiation.

Of course, these measures cannot prevent the damage that radiation can cause indefinitely, so crew members are only allowed to stay on the ISS for six months at a time. It's also why they must wear dosimeters at all times, which measure the amount of radiation they are exposed to regularly. To further protect themselves against radiation, crew members are given foods high in vitamins A and C. Such foods can produce free radicals, which can eliminate radiation in the human body. Foods that are high in fiber, called pectin and omega-3 fatty oils, are included in their diet as well because they are quite good at reducing the damage long-term radiation can do to the body. In addition, crew members are supplied with plenty of strawberries, kale, blueberries, and spinach (*Everyday Life on the ISS*, n.d.). This is because these food items prevent the neurological damage that radiation may inflict. Add to that the fact that crew members have plenty of supplements and medicines like radioprotectants, which can protect their cells against radiation, and it becomes clear that the ISS definitely takes cosmic radiation seriously.

All in all, Astronauts are given a menu featuring 300 different food items on the ISS. Most come in plastic wraps, except for anything liquid, of course. A lot of the food is freeze-dried to keep longer since making supply runs to the grocery store is hardly possible. Astronauts can add a dash of water to them and heat them in the oven in the ISS galley. In the 1960s, astronauts used to eat food from plastic tubes by sucking them out. Given their varied menu options, they must no doubt appreciate how far they've come since then.

Going back to microgravity for a bit, living in an environment where the effects of gravity are so low as to feel nonexistent has some very interesting effects on the human body. You see, on Earth, gravity is always exerting a pull or force, if you will, on the human body. The human body, in turn, actively fights against this force to stay upright. This doesn't happen in a microgravity environment like the ISS, though. So, with no such force to continually fight against, crew members' bones and muscles don't get the regular exercise they're used to. Thus, they start growing weaker, or they would if crew members weren't required to exercise daily while on the ISS. ISS crew members are expected to do strength, that is to say, weight training at the ISS to keep healthy and strong. They're also provided with treadmills to do cardio. As they run, they wear rubber straps that help keep them on the treadmills. In doing so, they keep their musculoskeletal structure strong and maintain their health to the best of their abilities.

Another interesting feature of microgravity is that it makes things float, as you know. This includes the human body. Hence, crew members must strap themselves down to the beds while sleeping, lest they float away a la Sabrina the Teenage Witch, bang against something, and injure themselves mid-dream. Some astronauts prefer using sleeping bags tethered to their beds rather than strapping themselves on their beds. They cannot avoid strapping themselves down on the toilets as they use them. The toilets at the ISS are very different than the ones on Earth. When crew members must use the toilet, they first strap themselves down on it. Once they're done, their waste is sucked away using a vacuum cleaner-esque system. Water, then, isn't used in these toilets.

Meanwhile, the toilet stalls don't have doors. Instead, they have curtains. This isn't as much of an issue as you'd think, though, because the noise of the toilet motors and air conditioning fans, which are needed to prevent bad smells since no one can crack open a window on the ISS, mask whatever sounds the toilet occupants make, anyway.

Regarding restrooms, crew members neither have bathtubs nor showers on the ISS. This is because the water they'd use would float away when they turned the shower on. Generally, crew members "wash" themselves clean using damp washcloths with body wash. They wash their hair with dry shampoo and wipe it off with dry towels. They similarly use wet towels or wipes to "wash" their hands, as there isn't a single sink on the ISS.

What about illness? Despite their best efforts to stay healthy, astronauts and cosmonauts could get sick while up in space. They're only human, after all. Recognizing this fact, crews always include a crew medical officer. Sadly, said medical officer is not named Dr. Leonard "Bones" McCoy, though who knows? Maybe one day, someone on the ISS will be given that nickname. In any case, this crew medical officer is the doctor on board, and they care for whoever gets sick on the ISS. The ISS is equipped with a full medical kit and a broad range of medicines. Just in case, all crew members are given basic medical training to know what to do if someone goes into cardiac arrest or how to suture a wound, for instance.

The crew members on the ISS aren't cut off from the world beneath them, at least not completely. They use a couple of methods to communicate with people on Earth. The Amateur Radio on the International Space Station (ARISS) is one of them, which you'll find out more about momentarily. This is essentially a two-way radio system. Individuals on Earth can send messages to crew members on the ISS using ARISS. Getting approval to send someone a message on the ISS via ARISS takes a couple of months, though. ARISS works because it uses radio waves transmitted via satellites in orbit. Radio waves are what crew members use to talk to Mission Control in NASA, too. For private discussion, though, crew members use the intercom either with individuals at NASA (or another space agency) or with their family members. The intercom allows for private conversation without transmitting it to the world.

ISS crew members naturally also have wifi access because, to quote the first *Doctor Strange* movie, they're not savages. Wifi was officially installed on the ISS in 2008. Since then, crew members have been able to go on the web freely, and they've habitually updated their followers on social media channels such as Twitter or "X," as it might now be called. More access points have been added to the station since 2008, meaning crew members now enjoy a fast connection, something they must appreciate when communicating with Earth. Communication between Earth and the ISS is pretty easy, at least for the most part. There's a bit of a lag, but it typically is only between two to four seconds long, maybe a little longer, depending on which position the ISS is at that given point in time.

Working in Space

ISS023E021088

Astronauts and cosmonauts don't just live on the ISS. It's also their main workplace during the time that they're there. So, what's working in space like? Astronauts spend a lot of time running various experiments and studies at the ISS. As of 2020, they had run 3,000 experiments on the ISS in total (*International Space Station Facts and Figures*, n.d.). The crews on board spend a total of 160 hours per week running experiments. They spend the rest of their time running maintenance tasks, doing space walks, running station control activities, or taking time to relax.

Crew members usually have two main tasks to take care of in the ISS: running experiments and doing maintenance operations. Overall, they do five different types of experiments, as you'll learn about in more detail in the coming chapters: educational experiments, experiments that are meant to make space safer for humankind, experiments to make astronauts healthier while living in space, farming research, and research to make new medical breakthroughs. Of these, the latter two may be a little surprising. I mean, why do farming and medical research in space?

There are several reasons for this. Farming or agricultural research is regularly done on the ISS because it helps crew members figure out how to farm in space properly. Figuring this out is vital if NASA and other space agencies are to succeed in embarking on longer missions, like the ones they're planning to Mars in the coming years since resupply missions will not be possible during them. Another reason why agricultural research is done on the ISS is that it can and has led to some incredibly useful inventions and discoveries. One of these was the Bio-KES, which is a device that converts carbon dioxide into water, something that could have very interesting implications in the age of global climate change (Toothman, 1970).

A final reason why agriculture is studied on the ISS is that simply put, it's good for the crew members' mental health. Crew members spend months on end on the ISS, which is a metal structure they can't leave on a whim. That can affect their mental health rather poorly once they get over the excitement and novelty of being and living in space. Plants, however, are known to have calming, pleasant effects on the human mind. Being around plants can significantly improve your mood and mental health, which the ISS definitely takes advantage of by having crew members conduct agricultural experiments (Haupt, 2023).

A lot of the agricultural experiments that are done in the ISS are all about improving agriculture back on Earth. One of these experiments has resulted in the development of the International Space Station Agricultural Camera (ISSAC), which takes infrared photos of farmland, forests, and similar spaces on Earth (Kim et al., 2012). These photos are then supplied to farmers, who use the information they've obtained to make better agricultural decisions.

As for medical experiments, these are often conducted on the ISS because microgravity, as will be explained in the coming chapters, affects bacteria in some very unique ways. Often, scientists obtain the results of their medical experiments much quicker in the microgravity environment of the ISS than they would on Earth. This leads to faster discoveries, innovations, and developments. A great example of this is the Recombinant Attenuated Salmonella Vaccine Experiment (RASV), which is all about finding a vaccine for pneumococcal pneumonia (Curtiss et al., 2010). This experiment is being done in space because microgravity speeds up the development of the bacterial processes involved in this disease. Scientists, therefore, believe that they will be able to find a solution to the disease more quickly by running experiments on it in space than on Earth.

Crew members conduct an array of educational experiments on the ISS, too. Most of these are conveyed to individuals, particularly students around the globe, using the ARISS. This program allows astronauts in space to communicate with astronauts, cosmonauts, and, more importantly, students on Earth (Campbell, 2017). Students get to talk to astronauts while they're on the ISS, ask them questions, and learn new things. The program counts as "scientific research" because it was born as an experiment to see if it would work, as you'll discover in Chapter 5.

Then, there are experiments that are about making spacecraft, including the ISS, safer for people. One of these experiments was the Microbial Growth Kinetics under Conditions of Microgravity (Huang et al., 2018). This experiment studies how microgravity and space flight affect bacterial growth. The study showed that the radiation in space and the microgravity affected the growth rate of bacteria. This was important data to have because it indicates that, depending on what kind of bacteria you're dealing with, the disease they cause can develop faster or slower in space. It also indicated that the speed at which antibiotics work on individuals may slow down or speed up in space as well. Studies like this allow astronauts to figure out what they and scientists on Earth need to do to make space a safer environment for people to be and live in.

So, crew members are allowed to spend six months at most on the ISS, as you know. This isn't just because of radiation but because being in space for extended periods of time can affect the human body in a variety of ways. It can even prove to be dangerous. While astronauts and scientists are running experiments to make conditions safer for mankind, this is a process. To understand why the experiments that are focused on making space safe for humans are important, we must first understand how space changes the human body.

For starters, some astronauts experience what's known as space sickness when they first head up to the ISS (Khalid et al., 2023). Space sickness is a condition that results from being in a lower gravity environment than that found on Earth. People who experience space sickness usually experience nausea, vomiting, and headaches. Luckily, space sickness goes away after you've gotten used to the new environment you're in, which can take a couple of days. When you return to Earth, you might experience the same symptoms. If you do, then you'll be experiencing gravity sickness, which will, again, take you a couple of days to get over.

Astronauts' faces swell while they're on the ISS. This is because, on Earth, gravity causes blood to head more toward your lower body. In space, less gravity means more blood travels to your upper body (*How the Human Body Changes in Space*, n.d.). Thus, fluids accumulate there, making your face swell—meanwhile, the mucous membranes in your nose also swell. So astronauts usually get congested in the ISS. After a few weeks, your body adjusts to the gravitational changes, so your facial swelling goes down.

Meanwhile, extended missions in space can cause an astronaut's eye shape to change. Specifically, it can make their eyes flatten a bit, change the quality of their vision, and cause edemas to form in their optical disk. Increased time in space can result in Space-Associated Neuro-ocular Syndrome (SANS), which is a kind of swelling in the back of the eyes (Yang et al., 2022). Scientists are actively working to solve this issue, given that planned missions to Mars will require that astronauts be in space for up to three years.

Astronauts sometimes experience a heart arrhythmia in space, too. They can also experience a decrease in their blood volume and aerobic capacity. Their heart and cardiovascular system still function well, though, since microgravity lessens the strain that is put on the cardiovascular system. This can result in the size of your heart shrinking, as well as make your conditioning levels drop over time. That can, in turn, mean coming back to Earth with a weaker heart that could overexert itself, which is another reason why you need to keep working out while in space. Then, there's the risk that the radiation in space might impact the endothelial cells lining the blood vessels. If it does, then this can increase your risk of developing coronary heart disease and speed up its progression. These are all problems that scientists and astronauts are actively working to solve.

Lastly, there's the way in which being in a confined space for extended periods of time can affect you. Granted, the ISS is quite a bit bigger than a regular spacecraft, but it's still a confined space that you can't leave for several months. Cracking open a window is out of the question, too. That can take its toll on your mental health and make you more stressed. This is one of the reasons why astronauts are given plenty of downtime. It's also why they're provided with spaces where they can talk to friends and family and why an extra effort is made to make sure space food is tasty and features all the culinary styles of the member countries that the astronauts hail from.

Experiments that focus on keeping astronauts healthy partly focus on these varying issues, as do some of the studies that try to make space safer for human beings. Additionally, health experiments take a closer look at how the various factors in space, like radiation and microgravity, affect their health. They further examine the way the health risks that might result from this can be mitigated. One such experiment was the Effects of EVA and Long-Term Exposure to Microgravity on Pulmonary Function (PuFF) study (SpaceRef Editor, 2002). PuFF investigated how microgravity affected the lungs. Before this, it was believed that the lungs were sensitive to gravity. This was proven to not be the case.

Cultural Exchange

Since the ISS is an international operation, it's home to many different individuals from many different cultures and belief system at any given point in time. Since international cooperation is a very important part of the ISS, cultural exchange is a cornerstone of it, too. There are many ways in which cultural exchange takes place on the ISS. Take food, for example. Ground crews make a genuine and thorough effort to include foods from diverse cultures in the astronaut's meal plans, especially those from member nations' cuisines (Crespo, 2023).

Providing crew members with meals from their own cultures is considered to be very important not just for cultural exchange but also to make astronauts and cosmonauts from other nations feel welcome and at home so far away from home. It lessens the stress they're under and any homesickness they may feel and improves their job performance. Astronauts from all nations enjoy the variety they're provided with, both because they so often prove to be delicious, improving the quality of the food that they eat, and because it allows them to try new things.

So, how do international crew members collaborate? For starters, there are certain legal frameworks to help them to do so. Different parts of the station were added on by different nations, as you'll recall. Whatever part a specific nation added to the ISS is considered their territory. Hence, that nation holds jurisdiction over that module and is legally responsible for it and the goings on inside it.

Chapter 4:

Scientific Breakthroughs

The building of the International Space Station is something wonderful, and it will show us how to take the next step beyond low-Earth orbit. — Gregory H. Johnson

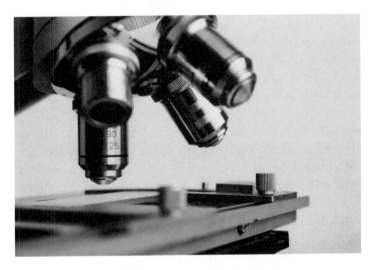

By now, it should be abundantly clear that a staggering number of experiments take place at the ISS. These experiments all focus on a variety of different areas and subjects. However, it seems some subjects get a little more attention than others. Medical research is one such subject, as you might have guessed from the previous chapter. So are Earth and environmental studies if the agricultural experiments we discussed are any indication. The same can be said for astronomical research, seeing as the ISS puts the researchers in the ideal position to study astronomical and space-related phenomena. What kinds of things do these various fields of study focus on in space, though? What interesting discoveries have they made, and what discoveries do they stand to make in the future? Let's find out.

Space Medicine

Many of the experiments conducted on the ISS focus on human health and a broad spectrum of medicine. To date, studies have been conducted to understand diseases like Alzheimer's, Parkinson's, various types of cancer, asthma, and even heart diseases better. These diseases are studied in space because, without gravity, the cells infected with them develop better and faster than on Earth, allowing scientists to understand their properties and behaviors better and develop new treatments for them (Shirah et al., 2023).

Several studies that were conducted in the field of space medicine, so to speak, focus on the effects space has on the human body. We know a great deal about how space changes the human body, but what we know is only the tip of the iceberg. The experiment that astronaut Scott Kelly participated in a couple of years back was enough to prove as much to us. Scott Kelly is an American astronaut and an identical twin. His twin brother, Mark Kelly, is an astronaut as well, or at least was before he turned to politics. One of the last missions that Mark Kelly participated in was an Earth-bound one. It was also a bit of an experiment. You see, NASA wanted to study the long-term effect of being in space on the human body. They said this was vital information to have if they were to succeed in conducting long-term missions in space.

To successfully observe how space changed a human being on a physiological and even genetic level, NASA devised a unique experiment involving the identical twins Scott and Mark Kelly. The experiment was quite simple. Scott would suit up, head to the ISS, and spend an entire year, double the time astronauts are traditionally allowed to stay in space. Meanwhile, Mark would remain on Earth and go about his daily life. The experiment served two purposes. The first was to see if the human body could be in space for that long. The second was to see if such a long time in space could change a person at the DNA level enough that observable differences would arise between identical twins.

As you can see, the experiment was quite long and contrary to some people's expectations, it was not cut short. Scott spent a year in space and then returned to Earth safe and sound. Upon his return, the identical twins underwent a series of tests to see how and if Scott's time in space had changed him. It had. For one, it had slowed down his aging. Human DNA has something called telomeres at its ends. You may think of them as DNA aglets if you'd like. Each time cell division occurs, those telomeres become shorter and shorter until they wear away completely. Thus, people age. Shorter telomeres aren't just associated with age-related diseases, though. They're associated with problems like Alzheimer's, too (Shukla, 2023). People who have early-onset Alzheimer's, for instance, typically have much shorter telomeres than their healthy counterparts. So, solve the telomere problem, and not only can you slow down aging itself, but you could also prevent diseases like Alzheimer's.

The scientists in charge of this twin experiment went into it thinking that Scott's telomeres would be shortened by his time in space, thanks to all that cosmic radiation. They were very surprised to find that that was not the case. Looking at Scott's telomeres and comparing them to how they used to be before he headed to space, as well as to Mark's telomeres, they saw that they hadn't gotten shorter. They hadn't even maintained their previous length. Instead, they'd gotten longer (Lewin, 2019). This wasn't the only surprising result to be observed in the study. The scientists also saw that Scott's telomeres became shorter again, as short as they used to be when he was on Earth full-time, once he returned to the planet and adjusted to living here.

Why on Earth—pun intended—did these two things happen, though? That right there is the million-dollar question, one that scientists are still trying to figure out. The current theory is that it has something to do with the combination of microgravity and cosmic radiation in some way. Cosmic radiation is also thought to be responsible for the changes to Scott's DNA. We all have certain genes in our DNA. Some of those genes are expressed, meaning they are active or online, if you will. Others are unexpressed and dormant. Scott's genes didn't change when he went to space. He didn't miraculously acquire new genes or lose the ones he had before. However, some of his previous genes, which were previously dormant, became expressed. Now, most of those genes have returned to their original state of being, whereby they were unexpressed after Scott returned to Earth. About 10% didn't. Instead, they remained "on" now that they had been switched on, including genes that control Scott's body's immune responses and DNA repair.

Again, why did this happen? Finding the answer to that question will take more research and experiments. In the meantime, the ISS will continue conducting other medical and health-related studies. Many R&D studies take place on the space station, examining things like aging, age-related diseases such as Alzheimer's, and muscle loss. Plenty of bacterial studies are also being conducted to come up with alternative treatments or cures for them. These experiments were of particular importance during the COVID-19 pandemic of 2020. For example, one study on the ISS focused on a drug called Remdesivir, an antiviral drug. It investigated whether Remdesivir could help patients suffering from COVID-19 (Cox et al., 2021). It was ultimately found that Remdesivir could be used to treat COVID, and this result was obtained much faster on the ISS than it would have been possible on Earth.

Medical research on the ISS has led to certain breakthroughs as well. One of these breakthroughs has to do with Alzheimer's. This breakthrough came from the use of a device called the Ring Sheared Drop device. This device can pin a drop of liquid, whatever it may be, between two rings (Guzman, 2022). It rotates one of these rings, thereby creating a difference in velocity between the layers of the liquid. When done with an amyloid plaque sample, the device can be used to develop drugs to remove this plaque from the human brain. Amyloid plaques are degenerative waste products whose presence in the brain is damaging enough to play a part in the development of Alzheimer's.

The amyloid experiment used the capabilities of microgravity, and the Ring Sheared Drop device made it possible for scientists to study Alzheimer's more closely and identify how it affects parts of the human brain more clearly. This data was then shared with pharmaceutical companies, who thus gained more detailed information to work with when researching drugs that could treat the condition. The study also showed that studying Alzheimer's in microgravity is easier because its processes work more quickly there, making it possible for astronauts to analyze it more quickly and thoroughly.

One medical condition that is often studied in the ISS is asthma. The ISS seems to pay special attention to the condition in its studies and respiration itself. One example is the ESA's Airway Monitoring study, which resulted in the creation of an instrument that measures nitric oxide levels in exhaled air (*Monitoring the Airways*, 2017). This instrument was effective in diagnosing inflamed lung conditions and asthma. Since its creation, hospitals worldwide have acquired the device in large quantities, and many were very thankful for it in 2020 when COVID-19 struck.

Yet another medical condition often studied on the ISS is cancer in all its various shapes and sizes. At the ISS, crew members usually cultivate stem cells to see how they might be used in therapeutic cancer interventions, especially with fewer and less intense side effects for patients (Grimm et al., 2022). Other cancer-related projects focus on 3D cell culturing, which is important for accurate drug testing and can improve the chances of effective drug treatments being developed for various cancers and the speed at which they can be discovered. One cancer-related project, for instance, developed a drug that targets cancer cells' blood supply, and only cancer cells are affected. In doing so, this drug deprives said cells of oxygen and nutrients, ensuring they die off. This drug was developed in the microgravity environment of the ISS. Developing this drug on Earth would not have been possible because the cells needed to run the necessary tests, which are called endothelial cells and help supply other cells with blood, die off very quickly outside the body under the ordinary gravitational conditions of the Earth.

Space medicine naturally also studies the human immune system. The thing is that astronauts' immune systems become impaired in space. Being aware of this, the ISS always conducts research focusing on how they can strengthen their immune systems (Dunham, 2023). Astronauts aren't the only individuals to struggle with compromised immune systems, though many individuals on Earth live with permanently impaired immune systems. So, experiments that focus on improving astronauts' immune systems in space, therefore, have the potential to help these individuals as well. Lucky for them, the studies conducted in space have revealed how immune systems become deregulated. These findings can now be used to figure out how this dysregulation can be fixed among astronauts and individuals with deregulated immune systems on Earth. While this has not yet resulted in a drug that can immediately be taken to fix the issue, with how things are going, it is only a matter of time.

A final medical breakthrough achieved on the ISS involves protecting muscles and bones, which, as you know, become weaker when astronauts stay in space for long periods, thanks to the lack of proper gravity. This is something that various diseases can cause back on Earth as well, and that can be a side effect of aging, as anyone suffering from osteoporosis would know. Hence, any treatment developed for this has the potential to help individuals struggling with such things. One discovery in this regard was that blocking a specific molecular signaling pathway helped protect astronauts' bone and muscle mass and strength (Lee et al., 2020). The study found that this measure can help individuals do the same on Earth. This breakthrough has prompted many pharmaceutical companies to turn their attention to drugs that could block this pathway already.

Earth and Environmental Studies

Of course, the ISS doesn't just conduct medical research, as necessary as that might be. The crew members on board also devote an equal amount of time to conducting Earth and environmental studies-related experiments and research. The most obvious way this is done is through observation, as the ISS does have a remarkable cupola overlooking the Earth. This may sound pretty basic, but the importance of observation cannot be understated, at least when done properly. Astronauts take considerable time observing the Earth or even enjoying the view. In the process, they record weather phenomena, like storms, and natural phenomena, like volcanic eruptions. They observe agricultural fields and have their observations relayed to farmers. Said farmers can then use this invaluable information to make or adjust their existing plans. Of course, the crew member's observations are paired with data obtained from sensors taking in all these phenomena.

To top that off, crew members often take photos of the Earth—in all honesty, who could resist?—and the resultant images prove invaluable for urban planning, disaster prevention, disaster relief, and proper resource allocation. Crew members also conduct the Earth's Surface Mineral Dust Source Investigation (EMIT). Mineral dust is important to keep track of because this dust can affect air quality, local temperatures, the rate at which snow melts, and the amount of plankton found in the oceans (Kramer, 2002). The study tracks the spectral signatures of the dust and investigates how the dust particles interact with the Earth's atmosphere, as well as the land and bodies of water in general. This naturally means that the study's results can have interesting implications for global climate change and its management.

Global climate change and global warming are things that the ISS focuses on especially. They've adopted the ECOsystem Spaceborne Thermal Radiometer Experiment on Space Station (ECOSTRESS), which takes infrared measurements of ground temperatures that show important differences and variations between urban and non-urban areas (*ECOSTRESS*, n.d.). Using this data, ECOTRESS identifies heat islands where extreme temperatures can occur. In the process, it seeks answers to important questions like "How is the biosphere—as in the plant life—responding to the changing landscape of water availability?" and "Can this advanced monitoring technology be used to reduce agricultural vulnerability and improve drought predictions?"

The SAGE-III might be a final example of how the ISS studies the Earth and conducts environmental studies. SAGE-III measures the various particles and gasses in the air, all of which play vital roles in the atmosphere and the natural processes that take place therein. Its predecessor, SAGE-II, helped establish a ban on certain harmful chemicals that ate away at the ozone layer in 1987. Today, SAGE-III can monitor about 70% of the planet and take measurements of the aerosols and gasses in its vast territory. For this, it uses occultation, the natural light that passes through the air, to take measurements. It uses this technique about 30 times daily (Flittner, 2017).

Astronomical Research

Lastly, the ISS conducts a fair bit of astronomical or space-related research. How could it not, given how ideally positioned it is? One of the major phenomena that the ISS studies in this area is neutron stars. A neutron star is the collapsed core of a supergiant. Aside from black holes, they are the densest and smallest known objects or entities in space, which is part of what makes them so fascinating to study. The ISS uses a special telescope called Neutron Star Interior Composition ExploreR (NICER), which can examine the formation of these stars following supernovae (Gendreau et al., 2012).

Since its launch in 2016, NICER has made some tremendous discoveries, including how X-rays spike on the surfaces of pulsars (*NICER (Neutron Star Interior Composition Explorer) TELESCOPE*, n.d.), and it has drawn maps of hot spots on pulsar surfaces. It has been discovered that pulsar cores are less squeezable than scientists originally thought. It has also made the most precise pulsar size measurements to date, which, given that you can't exactly walk up to a neutron star with a measuring tape is an impressive feat. On top of that, NICER has unveiled the first-ever pulsar surface map, caught a milestone x-ray burst, where the neutron star in question released as much power in 20 seconds as our sun is capable of releasing in 10 days, and has even managed to map the surrounding area of a black hole.

As you might have noticed, NICER is particularly good at catching X-rays. However, one piece of technology may rival it in this regard: Monitor of All-Sky X-ray Image (MAXI). The quaintly named MAXI was created and added to the ISS by the Japan Aerospace Exploration Agency (JAXA), one of the ISS partners, as you'll recall. Located on the Kibo module, MAXI was able to identify the source of the x-rays pretty quickly, albeit with the help of Penn State University's Swift telescope as well, discovering where its source was located in short time (*Space Telescopes Reveal Previously Unknown*, 2010). It's believed that this source is either a black hole or a neutron star, a couple of ten thousand lightyears away from the Earth but still in the Milky Way Galaxy.

This wasn't the only discovery that MAXI has made, not by a long shot. As you likely know, science fiction has long been fascinated with black holes. There have been many theories on how they work, in both science fiction and actual science, but it wasn't until MAXI came onto the scene that scientists figured this out. MAXI first observed the way black holes siphon gas from nearby companion stars. This has led to greater investigation into them and even fueled studies of antimatter and dark matter, which is vital for understanding the universe's composition (Mohon, 2020). MAXI achieved this feat when it caught a black hole outburst on the camera.

JAXA, it seems, is especially focused on astronomical research on the ISS. For instance, the space agency is searching for dark matter signatures across space. At the same time, CALET, JAXA's charged particle telescope, which can again be found on the Kibo module, scans space to discover the origins of cosmic rays, which are responsible for space radiation (*CALET: Experiment - International Space Station*, 2013). In the process, they aim to identify where cosmic rays specifically come from and discover what dark matter truly is.

Chapter 5:

Inspiring the Next Generation

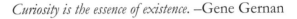

Curiosity is the essence of existence. –Gene Gernan

S121E06583

The work that the astronauts and cosmonauts aboard the ISS do is vital for the future, especially given the depth and breadth of their research. In a way, their work can be considered a sign of just how future-forward the ISS is. Yet this is not the only sign. Another key indication of this is the value those who are involved in the ISS, from the crew members to the ground crews around the world, are in educating the generations that are the future of space exploration. The ISS considers nurturing that generation to be an essential part of its mission. To that end, the ISS has not only a vast number of educational programs but also a variety of outreach projects. You never know, after all, how you're going to spark someone's interest or curiosity and light a fire under them.

Educational Outreach

Given the size and scope of the ISS, especially considering just how many countries and space agencies are a part of it, it shouldn't be surprising to find out that their educational outreach programs are not only numerous but also quite extensive, not to mention global. The space station has started many educational outreach programs throughout the years, and the opportunities these programs present are available all across the globe. Generally speaking, one does not need to be a citizen of a particular country to take advantage of them.

In all honesty, the ISS offers too many educational outreach programs for us to cover them all here. However, we can cover a select few. The ISS has many programs that foster their enthusiasm for space exploration and inspire the next generation of scientists, astronauts, and engineers. One of the programs they've launched to that end is ARISS. ARISS is the radio on the ISS, as you'll recall, and it's particularly useful for educational purposes. This is because students worldwide and of all age groups can use the radio to communicate with astronauts in real time and ask them any questions they may have. These questions may be about the ISS, the experiments run on the station, space itself, or any manner of thing they can think of. The goal here is to get students interested in science, technology, engineering, and math (STEM) subjects and in STEM careers early on. In this regard, NASA adheres to "the earlier, the better" logic.

Amateur radios were first used in space shuttles in 1983. They've come a long way since then, if ARISS is anything to go by. ARISS is the product of a partnership between NASA and the American

Radio Relay League, the Radio Amateur Satellite Corporation, various amateur radio organizations, and a number of the other space agencies involved in the ISS. The students who contact the crew members of the ISS do so from their schools. Every year, somewhere between 15,000 and 100,000 students make contact using ARISS (Love, 2023). Students can't contact the ISS crew at all hours of the day, willy-nilly. Instead, talks have to be scheduled in advance. Communication is established through actual radios and is supported by amateur radio operators and stations on the ISS. That being said, astronauts can make unscheduled contact with those wanting to contact them at their discretion.

ARISS isn't the only education program associated with the ISS. The ISS seems to favor one type of educational program, though: student competitions. The ISS regularly holds various STEM competitions among students. One of these competitions, which is open to middle and high schoolers, is called *Genes in Space* (Love, 2023a). The idea is to have students design experiments that use DNA analysis to solve actual problems ongoing in space exploration. The competition occurs in the fall and early spring, and students can participate individually and in teams. The experiments they design have to use polymerase chain reaction (PCR)—the method DNA replication occurs—analysis.

You might not think that students and children don't come up with anything worth noting in these competitions, but you'd be very wrong. Different students have made many scientific breakthroughs through the years through the competition. These breakthroughs have proven useful in scientific research and experiments and have even led to new developments, inventions, and innovations. In 2015, the winner, Anna-Sophia Boguraev, devised a new way to study how space flight affects immune cells. Astronauts on the ISS performed her experiment in 2016. Currently, her method is considered the go-to approach to studying immune cells where immune system experiments and studies on the ISS are concerned. That makes an abundance of sense when you consider how groundbreaking and how much of a success the initial, replicated experiment was. After all, Boguraev's method marked the first time PCR had ever been used in space. Meanwhile, the 2016 winner, Julian Rubinfien, came up with a new way of measuring the length of telomeres. Given how telomeres change in space, this new approach might be vital in future studies focusing on aging and diseases like Alzheimer's.

Not all of the ISS' educational programs are competition-based. Others are about keeping students updated on the most recent goings on the ISS. This is what the project *ISS Above* is about (Keeter, 2018). *ISS Above* is a small piece of hardware that can be installed in a classroom and that provides students with instant and current information about the ISS and where it is in orbit at that exact time. It's also a piece of technology that does HD live streams from the ISS. The screen flashes red whenever the space station passes over the students' current location. Keeping the station's existence and presence fresh on students' minds evokes students' curiosity and interest in the space station and in space exploration. The device is mostly used in middle school classrooms but can be installed in those of other age groups, too. It can be connected to a TV or projector to make things interesting, especially since the ISS' HD live streams show images of the Earth taken from the ISS.

Some ISS projects focus more on narrative and storytelling than anything else. That's not to say they're fictional, but they establish a specific audience-narrator relationship, with astronauts serving as narrators and students being the audience. A great example of such programs is *Science Time from Space* and *Story Time for Space*, which have astronauts on the ISS demonstrate various scientific concepts or read them story books from space—directly in front of the cupola overlooking the Earth, in fact—about space to students (*Science Time from Space*, 2020). The first of these programs' scientific concepts are typically very basic.

Experiments, therefore, focus on things like freefall and Earth observation. In any case, the two programs do this via live streams and recordings. The target audience is typically quite young, and this, of course, is intentional to get students interested in space and STEM at a very early age.

Finally, there's the *Space Station Ambassadors* program, a wide network of volunteers who share their knowledge and excitement about the ISS with students worldwide. These students can be in elementary, middle, or high school. The ambassadors themselves are typically educators, like science teachers. In other words, they're usually not individuals who have been to space or the ISS themselves. They are, however, significant advocates of both and work hard to get students interested in the two. To that end, they conduct workshops and seminars, develop new educational activities and learning opportunities, and advocate for further ISS research. There are about 2,000 ambassadors worldwide, though that number is ever-growing, and you might actually be able to become one if you want. All you'd have to do is apply for the position on the ISS National Laboratory's website (*Space Station Ambassador Program*, n.d.).

Media and Public Perception

There is an abundance of educational programs for kids and students of all ages, and the ones we've named are, again, just the tip of the iceberg. What about us adults, though? Does the ISS have any interesting programs for us, or is there a cutoff when we reach the age of 18? The short answer to that question is "no." We adults aren't left in the dark about the goings on and accomplishments of the ISS once we age. It's just that the ISS-preferred methods of getting us interested and engaged changed a little. Their approach becomes less experimentation and storytelling-based and more about utilizing the power of the media, social media, and the technologies available today.

For starters, regular media channels and news programs are fed a steady stream of information about the ISS. The media gets information about the ISS from the space agencies involved, such as NASA, JAXA, and ESA. Sometimes, these agencies provide live feeds of new ISS missions, making their audiences feel part of the process. These images and videos make space exploration feel much more real to many, which is a hard feat to accomplish, as the idea of going to space and staying in orbit is so general and foreign to most of us. This was the case for the new space station cargo launch docking mission on December 1st, 2023. The spacecraft carrying the cargo was the Russian Roscosmos Progress 86, which carried three tons of food, supplies, and fuel to the astronauts on the station (Malik, 2023). The docking mission was live-streamed on the web and made available on YouTube and NASA TV. Thus, the general public got a very good idea about how astronauts and cosmonauts aboard the ISS were kept alive and well while getting a sense of how international cooperation worked on the space station.

Now, you may be wondering what NASA TV is. NASA TV is NASA's official streaming and on-demand channel. Available on a broad range of platforms like Hulu, AppleTV, and Google Fiber, to give a few examples, it has been around since 2010 and is accessible to both the media and the general global public. It covers any new launch and features the daily activities of astronauts on the ISS, as well as footage of Earth-bound crews working with the space-bound ones (*NASA TV Live*, 2022). The schedule for its program is typically announced on NASA's website, making it easy for you to know when to tune in to watch specific shows. All this footage doesn't just remain online; it's made available to cable and satellite news programs. The video quality is HD, no matter where the footage is seen from.

Aside from an actual TV channel, the ISS has participated in its fair share of documentaries, having gone as far as to film things on the station itself. One such well-known documentary is *the Space Station Tour*, which gives its viewers a full tour of the station, walks them through its history, and shows them how everyday life works on the ISS (*International Space Station Tour*, 2012). The documentary features Sunita Williams, one of the most experienced spacewalkers in NASA's history. It records her final voyage as a commander for NASA, which took place in 2012 and covers her final 12 hours at the ISS.

Another documentary is *Moonrise on the ISS*, which records the astronauts daily lives on the space station and discusses the future of the ISS. Unlike most documentaries, it was filmed entirely by the astronauts themselves, meaning that astronauts had to learn the basics of cinematography and filming to accomplish this mission well. The astronauts it features are some of the candidates for the Artemis Program, which aims to put mankind back on the moon. One of these astronauts is Anne McClain, who dreams of becoming the first woman on the moon. The documentary focuses a fair bit on her but also devotes plenty of time to the ISS' other crew members. For example, it captures a historic moment, featuring Christina Koch and Jessica Meir performing the first all-female spacewalk in 2019 (*Moonrise from the ISS*, 2012).

Finally, there's PBS' *A Year in Space*, a docuseries covering the year Scott Kelly spent in space. It starts with Kelly launching into space and ends with his return. Nominated for a primetime

Emmy, it's 12 episodes long. It is pretty thrilling, given that one of the episodes covers a rocket explosion that generated space debris that the astronauts had to deal with (Kluger, n.d.).

The Role Models in Orbit

Speaking of Scott Kelly, there's no denying that he has achieved an incredible feat by going on and completing the yearlong mission he had been given on the ISS. After all, Kelly broke the record for longest stay in space when he went up to the ISS in 2016, though his record has since been beaten. Kelly stayed on the ISS for 340 days as part of an experiment to see if this was possible and doable and its repercussions. Given that, he can easily be considered a role model to look up to. Of course, the ISS has been a temporary home to many impressive role models. As covering all of them in one go is a bit of an impossibility, we have no choice but to focus on just a select few, starting with Kelly.

Scott Kelly was born in 1964. He became very interested in space exploration when he was a boy. The catalyst of that interest was watching the Apollo moon landing, naturally. He was only five years old at the time. 1986, he graduated from the State University of New York Maritime College. Then, he got his master's in aviation in 1996 from the University of Tennessee, Knoxville. While there as a test pilot, he applied to NASA's astronaut program, as most test pilots did then, and was accepted. His first space mission was to run a maintenance operation on the Hubble Space Telescope. For this mission, he and his crew spent eight days in space. He went on many more missions after that, including a construction mission on the ISS in 2007. He went on this mission, not knowing that one day he would call the ISS home for an entire year.

Scott Kelly officially began his first historic mission in October of 2010. On this mission, he served as commander and spent 159 days in space or, more specifically, on the ISS (Howell,

2019). While Kelly was on this mission, NASA and Roscosmos discussed launching a year-long mission for one astronaut and one cosmonaut to go on, respectively. As you know, the idea was to see how the human body would change in a microgravity environment and under the influence of cosmic radiation for that long. Kelly qualified for the experiment because he was a twin, because of his experience level, and because he had not yet met the lifetime radiation limit that NASA sets for all its astronauts.

After being officially chosen, Kelly launched into space in 2015. His and his Russian counterpart's first few months were rough as several supply ship deliveries failed, forcing them to ration their food. Once that was resolved, things went fairly smoothly. Kelly performed three spacewalks on his year-long mission and conducted an experiment that saved the zinnia flowers from extinction, among his many other tasks. He officially returned to Earth on March 2nd, 2016, having spent 340 days on the ISS, becoming, along with his Russian counterpart, the first man to spend so long in space.

Kelly isn't the only inspirational role model to ever set foot on the ISS. There are many more, like Sunita Williams. First selected to be an astronaut in 1998, Williams was born in 1965. A graduate of the U.S. Naval Academy, she also has a master's in Science of Engineering Management from the Florida Institute of Technology. Thus far, she has conducted two vital space missions, Expeditions 14/15 and 32/33 (*Sunita L. Williams*, n.d.). Before launching these missions, Williams received extensive training as an astronaut. She even worked in Moscow for a bit, participating in the next contributions Roscosmos would make to the ISS. She also lived in an underwater habitat called Aquarius for nine days as part of her training.

Williams was appointed Deputy Chief of the Astronaut Office following her first mission. Following this appointment, she became the flight engineer of Expedition 32 and the International Space Station Commander of Expedition 33. She spent 322 days in space for these two missions and is ranked second among female astronauts in the all-time U.S. endurance category.

A final ISS role model to look to might be Frank Rubio, who bested Scott Kelly's record as the longest resident in space. Born in LA, Rubio was selected to be an astronaut in 2017 (*Frank Rubio*, n.d.). Before that, he graduated from the U.S. Military Academy and earned his doctorate in medicine. Before that, though, he clocked 1,100 hours as a helicopter pilot, 600 spent in airborne combat in Afghanistan, Bosnia, and Iraq. A certified flight surgeon, he became a West Point Parachute Team member, which was poetically dubbed the Black Knights. In the process, he performed 650 dives.

At NASA, Rubio broke several records. For instance, he went on Expedition 68 2017 on the Soyuz MS-22, becoming the first U.S. astronaut to perform the longest single-duration space flight. He also stayed in space for 371 days, thus beating Kelly's record. He traversed 157,412,306 statute miles on this mission and witnessed 15 space flights to the ISS. He was present to greet incoming visitors every time. Like Kelly, he performed three spacewalks, which lasted around 21 hours in total.

Chapter 6:

Technological Marvels on the ISS

Technology is best when it brings people together. -Matt Mullenweg

There's no question that the ISS is a technological marvel in and of itself. So, it makes abundant sense that it would also be home to individual technological marvels. This is especially true because many of the modern technological developments we see today were born from space exploration, as you've already seen, thanks to the numerous experiments conducted on the station. Many of these technologies and inventions were initially developed to solve various problems in space exploration or to achieve new feats in the field. This is why the current era of space exploration is known as the Fourth Industrial Revolution.

Many technological developments we have seen throughout the years have occurred within the ISS or as part of ISS-run projects and initiatives. Space-based technology development has paved the way for advanced research and development (R&D) while maximizing how the ISS can be used. Essentially, the ISS has allowed for incredible technological developments in various fields, including robotics, hardware prototyping, electronics, and computing, to name a few examples. It will continue to do so for years, at least so long as it remains in use, in orbit, and in commission.

Advanced Systems and Machinery

The great thing about the technological advancements made in the ISS is that they inevitably trickle down to the Earth. Take the imaging and sensor technology on the ISS, for example. The sensor and imaging technology on the ISS are very powerful because they're used to observe the Earth and parts of space. One such sensor is the Lighting Imaging Sensor (LIS). The LIS was installed on the ISS in 2017 (Blakeslee et al., 2020). It can detect the orbital signatures of lightning storms both day and night and is more accurate than any other sensor to date. As you may imagine, it's a huge asset in climatology, lightning physics, atmospheric composition, and more.

Of course, the LIS isn't the only sensor on the ISS. Many other types have been added to the station over the years. Some of these, like the hyperspectral and thermal ones, are quite powerful in and of themselves. Since 2000, these sensors have captured over 6,000 images of the Earth (Kramer, 2015). Truthfully, there are hundreds of sensors on the space station.

You might not think so, but these machines' tasks are vital for ISS life. The sensors don't just monitor the Earth, after all. They continually monitor the environmental factors such as humidity, temperature, and pressure on the station, helping keep the crew members on board alive and alerting them should anything go wrong (Stenzel, 2016). They also monitor the concentrations of various gasses throughout the station, like oxygen, carbon dioxide, and ammonia. This allows crew members to fix issues when they arise and even makes it possible for them to evacuate in a timely fashion should that be required, though so far, it has not been.

The ISS also has remote satellite-based sensors that monitor the Earth, as you know. These are the various revisit locations they've observed before once every two weeks. That way, they get to capture different light conditions. One example of these sensors is the International Space Station Agricultural Camera (ISSAC), which helps farmers on the Earth. It was developed by students at the University of North Dakota and can be very useful in natural disasters, too. In 2011, for example, ISSAC recorded the flooding of the Souris River, which can be found near North Dakota (Stefanov, n.d.). The imagery and data obtained in this instance helped officials with disaster prevention by allowing them to take the proper measures against flooding. It also helped nearby residents evacuate if needed, and resident farmers protected and planned their crops accordingly.

Then, the Hyperspectral Imager for the Coastal Ocean (HICO) is on the Kibo module. As Japan's contribution to the ISS, the HICO collects information on water purity, bathymetry, and shore vegetation. You can imagine how vital the HICO can be in the age of climate change, especially given the impact it will have on plant life across the globe. The HICO is the first space-born sensor to focus onshore or coastal vegetation. Able to cover an area of 8000 km2 in size, it's an innovative prototype that was later integrated into other satellite systems after proving its worth on the ISS (Lucke et al., 2011).

Sustainability in Space

The imaging technology on the ISS is very advanced. Most, like HICO, are the first of their kind and the most advanced versions of their ilk, at least at the time of their installation on the ISS. Hence, they have the potential to be adapted for commercial use in a wide variety of industries. The same goes for ISS' energy efficiency systems. Energy efficiency and, therefore, sustainability are of the utmost import at the ISS. The station has to use its reserves sparingly, especially since supply runs are months apart at best and can be delayed. Take power, for instance. It takes great power to run the ISS, as you might have guessed. This doesn't mean that the station has power to spare, though. Instead, IT means that the station focuses especially on conserving its energy and using it as efficiently as possible.

The bacterial fuel cells on the station are one example of how the ISS does this. These cells use bacteria to transform carbohydrates into free electrons, which are then used to power parts of the station. The microgravity environment of the ISS speeds up this energy conversation process, given its effects on bacteria. This technology was first installed on the station in 2007 as an experiment (de Vet & Rutgers, 2007). It was unique in that it chemically transformed actual bacteria into carbohydrates. This produced free electrons, which were then harvested as an energy source. The technology proved very efficient because bacteria develop faster in a microgravity environment.

The bacterial fuel cells of the station fuel several parts of the ISS, but not all of them. This is why the station also has solar panels, as you know. These constitute the older fuel-energy systems of the ISS. Old doesn't translate to "not in use," especially with the newer solar arrays added to the station in recent years. Originally given a shelf-life of 15 years, the oldest panels are showing signs of degradation, though. That doesn't mean they'll suddenly stop working, but it does indicate they will become less and less efficient as time goes on. Hence, it's good that the ISS has those newer panels and its bacterial system. There is a reason why new ones regularly replace their old arrays through spacewalks.

You might have guessed as much by now, but one value the ISS places great importance on is sustainability. This is because of two reasons. First, the ISS astronauts must make do with limited resources. It's true that supplies are sent regularly to them, but they can't exactly step outside and go to the grocery store should they need or want something. This being the case, they cannot waste anything that they have, and they try to make the best use of it possible. The second reason is that many astronauts and scientists consider the ISS the best environment to practice sustainability and figure out how to implement it better on Earth itself.

Sustainability is also important for the ISS because the station actively monitors global climate change and the ecological and environmental changes that come with it. It has been doing so for years, as you've seen. Being in a position of such awareness, the astronauts on the ISS can neither ignore the data they obtain nor the troubling implications they bring. Instead, they use it to communicate essential information to Earth and work on projects and experiments that can help solve the observed issues.

The ISS' US National Lab is particularly devoted to testing out new sustainability-related technologies and has doubled down on its efforts to do so in recent years. As of the writing of this book, they're working on 30 different sustainability-related projects. One of these projects is the Orbital Sidekick (OSK), which detects hydrocarbon leaks in oil and gas installations. This makes it possible for people to detect such leaks early on and fix them quickly before they can cause any serious damage (Werner, 2023).

The Orbital Sidekick can do this thanks to its five satellites, though that number will be upped to six in 2024. The system's sixth satellite will be dubbed the Global Hyperspectral

Observation Satellite Constellation (GHOSt). This satellite will no doubt increase OSK's sustainability and imaging efforts by leaps and bounds, thus benefitting the many partners it has created projects and experiments with, including the Rochester Institute of Technology Center for Imaging Science, the US Air Force, the US Geological Survey, and the Oak Ridge National Laboratory.

Another important sustainability project is Cemsica for CO2 Capture. This bit of technology captures CO2 emissions made on Earth, like at industrial plants and combustion exhausts, at a much lower cost than would ordinarily have been possible. Making carbon measurements much more affordable is vital if plant operators are to lower their carbon emissions as they need and aim to.

The very interesting thing about Cemsica is that it uses nonporous membranes, whose development is aided by the station's microgravity environment, to measure carbon emissions. The company that came up with it, also named Cemsica, is having the system be tested on the ISS because this environment is thought to improve the level at which those membranes can perform while separating CO2 particles from the air and capturing them (Esen, 2020). By collaborating with the ISS, Cemsica hopes to reduce the cost of developing these membranes, thereby making the technology cheaper to procure and use. One feature that astronauts are testing out with it is whether photosynthesis can be incorporated into it. If it can, this will make the system much more efficient, and the implications it can have on global climate change and global warming can be massive.

The ISS is home to even more sustainability projects than this. Another rather fascinating project concerns Algeo. We often think that only trees can photosynthesize and, thus, turn carbon dioxide into oxygen. This is not the case. Algae have this capability, too. If this ability can be harvested, then algae can be used to tackle global carbon emissions and, so, climate change. The question is "how? and it's this question that the ISS is currently studying (Greene, 2019). This is why you can find entire algae pools at the space station, along with some light-seeking bacteria. These organisms are kept in something known as the photobioreactor, and their growth is actively studied by the astronauts staying on the ISS. The idea is to figure out what hastens algae growth and how they turn carbon dioxide into oxygen. If that can be figured out, then these processes can be triggered by scientists on Earth.

Communication and Navigation

Another highly advanced and efficient technology on the ISS is its communication system. Astronauts on the station are always connected to Mission Control, thanks to NASA's relay satellites. These satellites form a constellation of sorts. Data travels from the station to them, and the satellites then transmit data to the White Sands Complex in Las Cruces, New Mexico, as well as to the Guam Remote Ground Terminal. From there, the data is sent over to NASA. The same process unfolds when NASA sends a message to the ISS, except it flows in the opposite direction.

In 2023, this relay satellite system that NASA and the ISS use got a little flashier with the installation of the Integrated Laser Communications Relay Demonstration Low Earth Orbit User Modem and Amplifier Terminal (ILLUMA-T). This system uses laser signaling and communication. Infrared lights are sent along the route we've covered, thus ensuring that information can be sent to and from the station faster (Murphy & Schauer, 2023).

NASA's communication system has been updated several times, and understandably so. The upgrades to the system have increased the amount of data that NASA can download from the ISS. As of 2019, NASA could downlink 600 megabits of data per second from the station (Schauer, 2019). For reference, in 2016, they could only downlink 300 megabits of data per second, so you can see how quickly communication technology has been advancing over the years. These updates were crucial in light of NASA's plans to open the Lunar Gateway and launch longer missions, such as the one to Mars.

The ISS has a stellar navigation system, too. There are a couple of different navigation systems at hand. One is the ESA's Galileo, which uses GPS and was formally made part of the ISS in 2016, though the satellite launches began in 2011 (*What Is Galileo?* n.d.). Made up of 28 different satellites, the system makes it very easy for those on Earth to instantly determine the exact positioning of the ISS. The system's algorithm uses orbital information obtained from the various satellites. Those on Earth must know exactly where the ISS is positioned and where it will be in X number of days because otherwise, they could not plan missions for it. The spacecraft they send up to the ISS, bearing new crew members who are to switch with the old ones and supplies for everyone there, would never be able to make it to the station otherwise.

For example, in 2012, SpaceX had a mission to the ISS. The mission was unmanned, meaning no one could correct the course should the initial trajectory be incorrect. Hence, SpaceX had to know exactly where the ISS would be on the date and precise time they would be heading to it. In this instance, SpaceX's spacecraft used GPS to meet the ISS. This marked NASA's first time collaborating with SpaceX, so it was particularly important for the company that their mission succeed, as it would play a part in the two organizations' continued collaboration efforts. SpaceX's Falcon 9 rocket, Dragon, made it to the ISS without issues. In the process, it actively tested whether GPS could be used to run rendezvous with the ISS and found that it could.

> **FUN FACT:** Dragon, which took on SpaceX's 2012 mission and where GPS was tested, had an interesting payload on board: James Doohan's ashes. James Doohan was known for playing Scotty in the original Star Trek series from the 1960s, which means that he got to play an engineer traveling through space throughout his career and then going to space posthumously.

Chapter 7:

The Future of Space Stations

We're in the space exploration business, and the outer solar system is a wild, wooly place. We haven't explored it very well. —Alan Stern

As wondrous as the ISS may be, the truth is that it was never meant to last forever. Everyone who participated in creating the station, from the engineers who helped build it to the astronauts who stayed on it, knew that a time would come when the ISS would be taken out of commission. Originally, the ISS was supposed to have been taken out of orbit in 2024. Yet here we are in 2024, and the station remains up there. This is because the United States is intent on extending its lifespan for another six years and has been for some time. Why only six years? NASA and the United States know the ISS cannot last long.

Beyond the ISS

Six years may not seem like a long time, but a lot can be achieved within that timeframe when you think about it, especially if you have the budget, capabilities, and brainpower that the ISS has. The best way to achieve anything is to do it with intent and a plan. Luckily, NASA has one and has set some pretty specific goals for it (*International Space Station Transition Report*, 2022), these being:

- enabling deep space exploration
- conducting research that is beneficial for mankind
- fostering the establishment and development of the US commercial space industry
- enabling and leading international collaboration on all fronts
- continually inspiring humankind to greatness

The first of these, enabling space exploration, entails developing and testing environmental control and life support systems for future missions and turning the ISS into an analog for the future Mars mission. The second is just as much about understanding and responding to climate change as it is about developing beneficial technologies for humanity. Developments that contribute to mankind's understanding of quantum physics and atom optics, along with developing products that can prove life-saving and only be developed in microgravity environments, fall into this purview, too.

The third goal has much to do with NASA's and the world's future space station plans, which involve launching an array of commercial space stations later. On the other hand, enabling international collaboration is a goal that has long been at the very heart of the ISS. At the very least, it will remain close to NASA's heart in the future, as they will be collaborating with other nations' space agencies on missions. Artemis, a mission to get the first woman on the moon, is a key example. As for inspiring humankind, this can mean anything from raising awareness of the wonders and benefits of microgravity to conducting awe-inspiring experiments and exploring the world of STEM. The possibilities here are truly endless.

The ISS may only have another six years left in it, assuming measures aren't taken to extend its lifespan further, but clearly, it's set to achieve great things till then. Still, there will come a time when the world will have to say goodbye to the ISS and look to what's next. This is because, technologically speaking, the ISS won't last much longer than that and because of the rising tensions around the world. The fact that things have been extremely tense between the US and Russia since the start of the Ukrainian War is not exactly a secret, at least not to those watching the news. The same holds for Russia and the other nations that are a part of the ISS, including Canada, Japan, and various European nations. These tensions have only been rising in the wake of the sanctions that those nations have placed on Russia following the War. This is why, in 2022, the head of Roscosmos declared that Russia would leave the ISS in 2025 (*Russia Extends Cross Flight Programme*, 2023).

Initially, Roscosmos was aiming to leave the ISS in 2024. They have since agreed to move their extraction date, so to speak, to 2025, given the complicated logistics this involves. After all, Russia operates six modules that comprise the ISS. Among these modules is the ISS' propulsion system, which keeps the station going in orbit. Pulling out on short notice could have very problematic, if not catastrophic, consequences for the ISS, hence the new deadline of 2025.

New Frontiers

So, what exactly will happen once the ISS is de-orbited? The short answer to this question is that new space stations must be built. NASA has already commissioned Axiom Space to build a final module to be attached to the station. The idea is that when the time comes to take the ISS out of orbit, this module will be separated from the station. It will stay in orbit while the rest of the ISS returns to Earth. Thus, it will become the US' new space station. To that end, Axiom Space has been awarded a contract worth $140 million. They're not the only company NASA has commissioned for space stations. NASA has also awarded three other companies, Blue Origin, Nanorocks, and Northrop Grumman, contracts worth $130 million, $160 million, and $125.6 million to design and develop their own space stations.

If you're wondering why NASA wants multiple space stations in orbit, it's because that way, they can commission their services and have many different projects take place simultaneously. In other words, faced with two experiments of equal fascination, NASA won't have to choose between one or the other or schedule one for a later day. They'll just have to decide where to run those experiments. Should these plans come to fruition, we will likely enter a new age of commercial space stations, which will have interesting implications for space tourism. Who knows, maybe space hotels will even become a thing. These new stations would have interesting health implications, too, and may come to represent new institutions of health tourism. One study completed at the ISS found that eye implants, which improve people's vision, work better in microgravity environments than down on Earth (Guzman, 2022a). That being the case, it's plausible that people will eventually travel to space to get their eye implants rather than head to a hospital on Earth. They may do the same for other procedures, depending on what other health and space-related discoveries are made.

NASA's investment in private space stations begs the question: How many space stations will be in orbit? After all, the US isn't the only country interested in having its station up there. Arguably, all the countries that have been a part of the ISS to date could try to put their stations into orbit. A number of them are already planning to do so. Russia has already announced its plans to put its space station into orbit. In time, new countries might do the same. India, for instance, fully intends to do the same. Some countries already have their unique space stations in orbit. China's Tiangong is a key example of this.

The answer to this question is as yet unknown. However, it will become clear over time. The thing about building and launching a space station is that it's an expensive endeavor. It's also a dangerous one. Do one thing wrong, and you risk the lives of everyone living on board that station. So, there is and will be only a limited number of companies capable of launching a space station (Foust, 2023). On top of that, space agencies around the world, as well-funded as they are, will only have so much money to pour into companies willing to build new stations. How many stations will ultimately end up around the Earth's orbit will depend wholly on supply and demand.

NASA has other exciting plans aside from its new commercial space station deals. One is to build the Lunar Gateway, a permanent space station in the moon's orbit with a perpetual human presence from now on. This space station will be much smaller than the ISS, and it'll be built with both governmental and private partners. The station is expected to launch in 2025. The Lunar Gateway will serve a variety of purposes. First, it'll serve as a pit stop for astronauts on their way to Mars since NASA intends to head to Mars in 2030. Second, it'll support all lunar missions, of which there'll be plenty. The first of those missions intends to put a woman on the moon for the first time, though the astronaut who will hold this honor hasn't been chosen yet. Third, it'll monitor deep space and serve as a waypoint for future deep space missions since NASA has no intention of stopping its space exploration efforts after reaching Mars.

Now, the Lunar Gateway won't just be NASA's doing. Other space organizations will contribute to it as well. The ESA, for instance, is adding an international habitat (I-Hab) to it. The I-Hab will have quarters for visiting astronauts, multiple docking ports, and several other modules. Four astronauts can stay in it for up to 90 days at a given time. The I-Hab will include labs for experimentation, too, and an attachment for the Canadarm-3 robotic arm, which, as you'll recall, was made for the ISS. Japan's JAXA will also contribute to the Gateway by adding its Environmental Control Support System and various batteries and cooling devices.

So, how will the ISS be taken out of orbit? First, NASA and other space agencies will let the ISS begin decaying rather than continually repairing and updating it. They will keep up with repair and updating measures until 2026. As a result of this, the station will drop from a height of 250 miles to 200 miles above the Earth (O'Callaghan, 2023). This will happen in the mid-2030s. When the drop happens, a crew will be sent to the ISS to remove any remaining important equipment, as well as items and equipment of historical importance, from the station. Once that's done, the station will drop to 175 miles above the Earth. This will be the point of no return in that, once the station drops to this height, changing their minds and boosting the station back to low orbit will no longer be physically possible.

When the station drops to a height of 75 miles above the Earth, it will hit the planet's dense atmosphere. When it does, its solar panels will be torn off because of the force of the winds it will experience on re-entry. At 50 miles above the Earth, the modules making up the station will start being ripped apart. Then, they'll start burning away due to the friction they'll experience on re-entry. The temperature generated from that friction will be around a couple thousand degrees. Hence, those burning modules will disintegrate, and loud sonic booms will echo across the sky as they do so. Any debris that survives the re-entry will fall into Point Nemo, a part of the Pacific Ocean between New Zealand and South America. This is a spot that has become a space graveyard over the years. There's very little marine life here, so it shouldn't be dangerous to animals there, either. To eliminate any risk of casualties, though, this portion of the Pacific Ocean will be restricted and out of bounds when debris is expected to fall.

However, some believe having the ISS land like this would waste valuable resources and materials (Foust, 2023a). Some have even started a change.org petition to land the ISS safely and recycle it. Among them are CisLunar Industries and Astroscale, which proposed reusing those materials, especially solar panels, instead of wasting them. They've brought this proposal to Congress in 2022. If the proposal is acted on, a sort of salvage yard can be found in space, creating a new way of dealing with spacecraft that are taken out of commission.

Challenges and Opportunities

NASA's plans to develop commercial space stations have certain technical and financial challenges. One of the technical challenges is that the space stations need specialized materials to be built to withstand the harsh conditions of space, such as cosmic radiation. Without these special materials, astronauts' lives and well-being can be at risk. Finding these materials can be challenging, though. Ensuring that the space stations remain intact as they're taken to orbit and deployed can be challenging, too, since the materials they're made with both need to be sturdy enough and constructed properly. No mistakes can be made in this regard. No one would want another Challenger disaster on their hands, after all, even with space stations.

Another technical challenge is developing reliable—that is to say, reliable in the long-term—life support systems for the space stations in question. While the ISS has done an admirable job of this, they've proven that their systems need to be regularly maintained and updated to keep functioning well (Russell & Klaus, 2007). The only possible way is through space walks, which astronauts must regularly perform. The same goes for any necessary emergency systems and the establishment of foolproof safeguard measures to protect the lives and well-being of the astronauts on these stations.

Docking and resupply are yet more challenges that must be figured out. A universal docking system must be established, uniform across all stations, and one that works with all spacecraft. Otherwise, different spacecraft won't work with specific stations. They won't be able to dock them and won't be able to deliver new crew members and supplies while ferrying departing crew members back to Earth. Developing a universal docking system, though, is a tall order. However you look at it, it will require much logistical planning and some degree of collaboration between different space agencies and companies. That's not always the easiest thing to do. Those are just the technical challenges. There are also financial ones, like the initial costs for building and launching the stations, which will be quite high. The companies contracted for this must invest heavily in research and development, manufacturing, and launches. Cutting corners will be, or at least should be, out of the question as that can cost lives and end in disaster.

There will also be ongoing operational expenses, which will be pretty high in and of themselves. Such expenses can quickly add up and surpass the budgets allocated to the space stations (Crane et al., 2023). If that happens, running the space stations, keeping them operational, and maintaining them properly will be impossible. This will either mean those space stations will be unable to run at optimal levels of efficiency or that they will decay more quickly, which can, again, have catastrophic results for all involved.

Two other financial challenges are figuring out how to generate revenue and managing the competition aspect of the space stations since there will be multiple ones in orbit now. Of these challenges, the former will be easier to solve since plenty of organizations and companies will be interested in having their experiments and research done on the space stations. This will become a source of income for the companies and organizations who created those space stations in the first place. The second challenge may be more difficult to manage since many of these stations will compete against one another to secure contracts, just as they would in an ordinary, Earth-bound free market. However, competition will be fiercer and much more expensive. Since the market will be small, at least initially, those who don't get a contract might not be able to find any other contracts or deals. The cost of a space station maintaining its competitive edge can strain things, especially in light of the expenses space stations will already be generating. Ultimately, this situation could result in a space station being removed from commission.

It is vital that challenges be solved before those space stations are launched and a transition is made from the ISS to multiple stations. Otherwise, research on microgravity and research involving microgravity can ground to a halt. Also, NASA's launch dates to the moon, Mars, and beyond could be delayed since important microgravity tests needed to ensure these missions can take place safely won't be getting done. Likewise, international cooperation could wither further, and the commercial space industry will take a hit, costing everyone valuable development opportunities in space exploration and tourism.

The competitive nature of the emerging space station market, because it is a market, presents both a challenge and an opportunity, just as any free market would. The difference is that it's a more expensive challenge and opportunity than most competitive markets. Take the space station that Axiom will be building. This station, which will be dubbed Haven-1, is estimated to be worth $2.4 billion (*Space Startup Vast Partners with SpaceX,* 2023). NASA may have contracted a select group of companies to build and operate space stations, but more companies are expected to join the fray as time passes. One startup, Gravitics, already has designs for a station called StarMax. Another startup named Above Space has already secured a space contract worth $1.7 million. The various space stations currently being developed and designed vary in terms of how they're envisioned and what technical approaches are adopted in their design process. For instance, Axiom Space aims to convert a shuttle-era multipurpose logistics module into a space station module and research lab. Meanwhile, Sierra Space is developing inflatable modules, a technology first demonstrated on the ISS.

The competitive edge to this market is as challenging as it is an opportunity because, at present, it's very hard to see how big the commercial station market will be and how many businesses can close their business cases. Most of these companies will need non-NASA businesses to close their cases, and they may encounter a lack of demand in this regard. Depending on demand, it may turn out that there isn't much of a market.

The emerging scene is indicative of one thing: the role of space agencies like NASA is actively changing and evolving in our current time with the emergence of new markets like space tourism and, potentially, space stations. Between 2000 and 2016, more than 350 new private, space-related businesses have opened up, and once they open, they grow and expand at a staggering rate. Currently, the focus of the space age is moving from global collaboration, as demonstrated by the closing of the ISS, to commercial participation and innovation. That's not to say global collaboration and partnerships won't still exist—if they did, the ESA and JAXA wouldn't have gotten involved in the Lunar Gateway—but it does indicate that that won't be its defining feature.

Chapter 8:

Risks and Challenges of Living in Space

Anyone who sits on top of the largest hydrogen-oxygen fueled system in the world; knowing they're going to light the bottom—and doesn't get a little worried—does not fully understand the situation. –John Young

Going to space and staying on a space station like the ISS is an exciting idea for most people. Most people would clamor for the opportunity to do so and start packing immediately. For some people, though, it's pure madness. This is because going to space isn't exactly child's play. Going to comes with a whole lot of dangers and risks. The same goes for actually staying in space for any measure of time. You've already been acquainted with several of these dangers, like cosmic radiation and the muscle and bone mass loss that astronauts experience in space. Others, though, may be wholly unfamiliar to you. As such, they may be entirely unexpected and different from your thoughts.

Space Dangers

One of the first risks that come to mind when considering living in space is space debris and things like asteroids. In other words, it's the risk of collision. We're used to thinking of space as emptiness, and it is. However, that emptiness contains a whole host of objects, both artificial and not, that a space station or spacecraft, for that matter, could collide with. Space stations are more likely to collide with space debris than satellites because they have fixed orbits and trajectories. Space debris is junk that has snapped off degrading bits of space objects, like satellites. They pose a very real threat to the satellites we have in orbit, as well as to the ISS.

When the ISS was first designed, the engineers working on it gave it maneuvering capabilities. In other words, they ensured that the ISS could perform evasive maneuvers. Since its establishment, the ISS has had to perform these maneuvers and thus avoid collision with space junk and micro asteroids 32 times (Howell, 2023). There have even been times when such collisions occurred, though thankfully, none of the astronauts and cosmonauts on the ISS were harmed. However, parts of the ISS were damaged due to them. Canadarm2, for instance, has suffered such a collision, as has the wraparound cupola window that faces the Earth (Datta, 2021).

Space debris from space junk is an ever-growing problem, given the number of satellites launched into space. To give you an idea of its scope, there are currently 5,465 active satellites in orbit. The trouble is that most of these aren't regularly maintained by astronauts performing spacewalks. So, they wither and decay over time, and at some point, bits and pieces fall and snap off of them. Space debris is a huge issue because "dead" pieces of satellites and other debris remain in their orbits, continually going round and round at the speed at which they were moving. They keep going like this until they crash into something and thus come to a stop. Another reason they're such a major issue is that there's no real or feasible plan to physically clean them up.

If you think that 5,465 is a lot, that's nothing compared to what will come next. SpaceX alone is planning to place 12,000 broadband satellites in orbit and has applied for approval to get an additional 30,000 up there. Luckily, SpaceX is designing its satellites to automatically come with a collision course avoidance system, allowing them to maneuver around at the first sign of a threat. There have been instances when space debris has necessitated that astronauts on the ISS take shelter, perform the ISS' evasive maneuvers, and cancel planned spacewalks.

Maneuvering a spacecraft as large as the ISS is no easy task. The ISS policy on evasive maneuvers is that these maneuvers have to be performed if space debris comes within a pizza box distance of the station, with the station being at the center of said box. If "pizza box" sounds too vague, the box has to be 2.5 by 30 by 30 miles. Typically, objects must be about 2 inches big for such maneuvers to be performed. Still, they can be done even when things as small as paint flecks are headed toward the station, provided they're moving at a high enough velocity. If they do, they can be as damaging to a space station as if they were actual bullets. In other words, they puncture holes into it, and this can lead to critical issues.

When a space object seems to be getting close enough to the ISS, Mission Control in Houston and Moscow contact the astronauts and cosmonauts at the station to give them their directives. The crew then takes the helm. So, the ISS gets out of the way of oncoming debris by firing up its thrusters and adjusting the station's orbit. This takes quite a bit of energy, though. Knowing this, the crew always keeps extra fuel on board if they have to maneuver. Without extra fuel, performing those evasive maneuvers wouldn't be possible. One example of such maneuvers occurred on October 25th, 2022. This was when Russian controllers fired up the thrusters of the Progress 81 cargo ship, which was attached to the ISS at the time, for five minutes and five seconds to avoid incoming debris. This maneuver raised the ISS's altitude by 0.2 miles at its summit, enough to avoid said debris (Sidharth, 2023).

Another obvious danger in space and at the ISS is radiation, as you know. There's much radiation in space, but the ISS isn't entirely defenseless against it. It's made of protective materials that are meant to shield against radiation. This protective shielding lowers the danger that comes with cosmic radiation. So does the food astronauts eat, as they include many ingredients that lower the effects radiation has on the human body. Still, radiation poses a danger to humans, so astronauts only stay in space for limited periods.

Radiation causes human cells to wither by breaking down the DNA strands inside them, which leads to long-term health consequences like cancer (*Radiation and Health*, 2023). Radiation knocks the bases that the human DNA, like guanine and thymine, is made of. When this happens, cells try to repair the damaged DNA. Sometimes, it succeeds. Other times, it does not. A cell can easily miss a gene or misrepair it, causing mutations that sadly don't look like the kind the *Fantastic Four* experienced after being exposed to radiation in space. In addition, radiation damages the human heart and cardiovascular system, narrows arteries, and reduces or eliminates the number of cells that line your blood vessels. This, in turn, results in various types of heart and cardiovascular diseases. Being aware of this and intending to launch long missions in space, NASA is developing new ways of protecting humans against radiation. As for what their studies will unveil and lead to, only time will tell.

Another major risk where space living is concerned is illness. No one wants the astronauts to get sick in space because it just spells trouble. Don't get me wrong; the ISS does not lack any of the resources it needs to tend to its crew members' health. That's why it's equipped with a medical officer and all sorts of medication besides. It's also why all crew members know basic first aid. Despite all this, though, the fact is that getting sick on the ISS is not the same as getting sick on Earth. On Earth, you can quickly go to the ER and get whatever treatment or surgery you need. In the ISS, this isn't necessarily the case. You might get some treatment on board, but if something seriously wrong with you can't be solved there, you'll have to be brought back to Earth.

That won't get done immediately. Arranging it will take time. Once everything is arranged, getting a sick astronaut back to Earth will be hard, especially for the astronaut in question. Imagine going through the turbulence of entering back into the atmosphere while your body shivers with a high fever and feeling nauseous, all while wearing a space suit you can't get out of.

Again, NASA knows how troublesome getting a sick astronaut home would be. So, NASA scientists have developed over 800 health standards for astronauts who want to travel to space (Sohn, 2021). These standards cover everything from what the astronaut's muscle and bone density should be, given the potential muscle and bone loss they can experience if they don't work out daily, to what their eyesight should be like, seeing as being in space can affect your eyesight badly. By choosing astronauts who adhere to these standards, NASA lowers their risk of getting sick to a minimum.

A final risk that astronauts have to prepare for before heading to the ISS is muscle and bone mass loss. This potential loss is why astronauts undergo physical training while on the ISS, as you know. They also undergo rigorous physical training to increase their endurance levels. This increases their bone and muscle mass and gets them strong enough to deal with any loss in mass that they might experience. Of course, astronauts receive intense psychological and isolation training, too. This is a necessity because they will be cut off from Earth, physically speaking, and cooped up in a relatively small space for months on end, and this can understandably take a toll on them. One of the ways astronauts train for this is by doing wilderness survival activities. One such activity has astronauts being dropped off in the wilderness at −13 F (-25 C) weather. There, they are only given a couple of key items, like matches, to survive. This prepares them mentally for the hardships and psychological isolation they'll experience in space. Likewise, it prepares them for any intense emergencies they may experience on the ISS. Though no major emergency, like one that requires an evacuation or a medical emergency, has happened on the ISS yet, it always helps to be prepared.

What if the crew on the ISS couldn't perform the necessary maneuvers in time, and the collision was inevitable? This can happen when the crew isn't given timely notice or if debris tracking proves even a little imprecise. Should such an incident happen, the crew is expected to take shelter in their escape vehicles and wait for the collision. In an emergency, like debris puncturing through the station, causing air loss and pressure, the crew must use their designated escape vehicles to return to Earth. Their escape vehicles typically consist of the Soyuz or SpaceX Crew Dragon. When taking shelter, they have to wear their spacesuits just in case. The crew had to take shelter like this in 2021 when a piece of debris flew too close to the ISS. Luckily, it didn't strike the station (Bartels, 2021).

In recent memory, there has been one significant collision: a micrometeoroid—which usually weighs 0.04 ounces and is no more than 0.04 wide and long—struck the ISS. The collision caused a devastating leak on the Soyuz, which was docked at the station then. This happened in 2012, and the leak was a small air leak. As a result of the incident, the spacewalks scheduled for that day were called off (Azriel, 2012). Luckily, it was determined that the crew was not in danger because of the leak, and they could fix the issue relatively quickly.

You already know what the emergency protocols for debris-collision situations are like. How about medical emergencies? After all, even with the best safety protocols in place, such an emergency may happen. One way of handling such an emergency might be telemedicine, which, while limited, is often used in medical emergencies in aviation these days (Wilke et al., 1999). This would be a good option to consider because handling the g-forces that the human body would be subject to on re-entry to the Earth would be very difficult for a sick person to do. Another long-term solution is to make full-time doctors part of every crew traveling to space, not just the ISS. Granted, performing surgery in space isn't a real possibility, even if the patient needs it. This is because the space station's microgravity environment would make blood leak out of the patient's body at the first incision. Said blood would then float about the station itself, potentially causing a lot of blood loss and infecting other crew members in the process.

Making it possible to perform surgery in space is something US scientists have been testing out for some time now. One of the options they're exploring is putting transparent domes over open wounds or cuts and filling them with saline solutions to stem blood flow in microgravity zones. Such a measure, should it prove effective, could give a surgeon the time to suture whatever wound they're dealing with. Another option NASA explores is robot surgeons (*About Robonaut*, n.d.). The ISS already has a robot on board called Robonaut, and the goal is to get it to the level that it can perform basic medical tasks while being controlled by Mission Control back on Earth. Should Robonaut become advanced enough, ideally, it'd be able to perform complicated surgeries, which can come in handy on extended missions like the one to Mars.

Chapter 9:

The Unseen ISS

Sorry. Just paralyzed by the indescribable beauty of the cosmos. We'll get to work. –Daniel Suarez

The astronauts and cosmonauts living on the ISS are undoubtedly heroic and awe-inspiring. However, they are supported by a whole host of individuals who work day and night to ensure their safety, well-being, and the success of their missions, even if they do so from Earth. Back on Mission Control, for instance, there is an individual whose sole job is to monitor the mountain of health data they get about every astronaut on the ISS every day to keep track of their mental and physical health and intervene if and when it's necessary (*Medical Operations Team Activities*, n.d.). So far-thankfully-it has not been necessary. So, who exactly are these unseen heroes behind the scenes? What precisely do they do? Let's find out.

Behind the Scenes

There is an entire group of people whose feet are planted on Earth and whose main job is to safeguard the lives and safety of the crew members on the ISS. Their ability to do their job depends on one crucial factor: communication. Mission Control must be able to communicate with the ISS at all times to do its job well. The further you travel out to space, the longer it takes to send and receive messages. For instance, if Mission Control wanted to send a message to Mars, it would take 20 minutes for that message to get there, at least with our current technology (Platt, 2020). Luckily, the ISS is in low orbit, so messages quickly arrive. Accidents still sometimes happen, though. For instance, In 2023, NASA briefly lost contact with the ISS due to a power outage, causing some panic among Mission Control members.

This was the first time a power outage ever happened on the ISS. Fortunately, the station has backup control systems, which it was able to switch to, marking the first time the ISS ever had to use this system, too. As a result of the outage, Mission Control briefly lost all control of command, voice communications, and telemetry with the station. About 20 minutes in, Mission Control was able to alert the crew on the ISS about what was going on using the Russian communication systems on the station. A statement was quickly released, explaining that the astronauts were not in danger at this time, and neither was the station itself.

The space station program manager is the person in charge of overseeing the management, development, integration, and operation of the ISS Joel Montalbano. On any given day, Montalbano had a lot of responsibility. How could he not when the ISS program costs nearly $3B annually (*Joel Montalbano, International Space Station Program Manager*, n.d.)? How could he not when he was in charge of all policy development, international partner negotiations, development of low-Earth orbit commercialization, onboard research and utilization, and the overall safety and health of the crew on the ISS?

Suffice it to say that Montalbano had a lot of tasks he had to tend to. The ISS power outage became his number one priority, though, when it happened. The same applied to everyone else working on the ISS. That's why Mission Control stayed overnight at the base the day the outage happened. They were aware, after all, that had things turned into an emergency—had the air conditioning and lights on the ISS shut off as well—Mission Control would have moved to a backup control center that was a few miles from Houston, a standard protocol that they haven't had to use to date (The Guardian Staff, 2023).

Much to their credit, Mission Control restored power and communication fairly quickly when this power outage incident happened. Thus, they could return to their regular responsibilities, like figuring out how to make the next supply run to the ISS. Regularly getting supplies to the ISS is one of Mission Control's and the ground crew's top responsibilities. However, the unfortunate reality about supply vehicles is that they can fail to launch, and some have done so in the past. Because of this, ground crews always try to keep extra supply vehicles and spacecraft at hand. This is also why the ISS crew values sustainability so much and strives to save any resources they can (Garcia, 2018).

Astronauts live according to tight schedules to complete their work on time and for their health. For instance, they must get at least 8.5 hours of sleep daily. Ground crews and Mission Control know their schedules and current supplies very well and try to plan resupply missions according to this information. To that end, they schedule daily planning conferences twice a day. This allows the ISS crew and Mission Control to catch up with each other and ensure everyone is on the same page and schedule at all times. Additional conferences are scheduled leading up to resupply and space walk missions.

A full-time job is figuring out how and where to store all the supplies and equipment sent to the ISS. As such, it's the responsibility of Inventory Stowage Officers (ISOs) in Houston and their counterparts in Tsukuba, Munich, Huntsville, and Moscow. These officers barcode the supplies to keep careful track of them. They work closely with flight control teams to build stowage notes for crew activities, which outline where the crew can find the necessary tools, equipment, etc. In addition, ISOs figure out how to unpack and put away cargo in arriving vehicles, how to pack whatever needs to be returned to Earth, and where to temporarily store items that will be thrown away. Despite all their diligent efforts, items sometimes get misplaced, leading to "wanted" posters being created for them. Every year, about eight or nine resupply missions are made to the ISS, meaning such a mission typically occurs once every 40 to 45 days (Evans & Laufer, n.d.).

Cultural and Personal Life

Astronauts live on tight schedules, as we said. Astronauts typically take Sundays off, though. No one can work 24/7, after all. They also take time off for various vacations and holidays, including national ones. This time off is naturally scheduled ahead of time. Take Thanksgiving as an example. The first space station crew to celebrate Thanksgiving was Gerald P. Carr, Edward G. Gibson, and William R. Pogue, who were on Skylab 4. This was in 1978 (Lea, 2023). The first group of astronauts who celebrated Thanksgiving on the ISS did so with Russian cosmonauts in 1996. That day, they watched the Earth together rather than watching football the way they ordinarily would have on Earth. They ate smoked turkey, freeze-dried mashed potatoes, peas, and milk for Thanksgiving dinner. In 2009, the largest international crew ever celebrated Thanksgiving together. There were 12 crew members, hailing from the United States, Russia, Belgium, and Canada. The crew celebrated Thanksgiving two days early because a shuttle from the ISS was to undock on Turkey Day.

Astronauts celebrate other holidays, too, like Christmas. Typically, they have a zero-gravity Christmas tree for this. This is an artificial tree with Velcro straps attached to it that keeps it from floating away. Astronauts usually decorate it with lightweight ornaments and keep it by the crew quarters. Christmas favorites like turkey, green beans, and cranberry sauce are freeze-dried and taken out on the day of. Crew members naturally place video calls to their families back home on Christmas day (Malik, 2023b). They also keep Christmas stockings in the ISS. These stockings are usually filled with letters from home, snacks, and personalized gifts. All astronauts open their stockings together to celebrate.

All ISS crew members share their different Christmas and other holiday traditions with one another. They also celebrate holidays that are specific to different countries or faiths together. On January 7th, for instance, they observe Russian Orthodox Christmas (Uri, 2020). They also observe and celebrate Hannukah, which lasts eight days. The first crew to spend Christmas on the ISS were William M. Shepherd, Yuri P. Gidzenko, and Sergei K. Krikalev. They recorded a goodwill message to all the people of the Earth. As for New Year, Shephard wrote a poem on the first day of the first year in the ship's log, a very old naval tradition.

Holding these celebrations and spending time like this together is very important for various reasons. First, it makes the crew on the ISS feel more connected to their cultures and homes. This helps with any isolation or homesickness they may struggle with. At the same time, it allows for greater cultural exchange among crew members from different cultures and further emphasizes global unity. Most importantly, it brings separate crew members together, making them feel like a team or family. This is more important than you might think because it affects the crew member's ability to work well together, whether they realize it or not (Guzman, 2022a). Crew members have to be able to "read" each other's emotions, and this ability of theirs may become impaired over time, given the psychological toll extended stays in space can take. If this is true for several months' long stays on the ISS, imagine how true it'll be for much longer missions, like the ones to Mars, since a long trip to Mars will take roughly three years. By figuring out ways to manage these psychological factors, astronauts and Mission Control can develop ways of looking after their mental health and preparing for those longer missions in the future.

Given all that, a scientifically approved mental health checklist is given to all the crew members traveling to the ISS (Lewis, 2023). Crew members should keep careful track of the items on this list and let their coworkers and those at Mission Control know if and when they're struggling with something. Incidentally, this list is also provided to scientists who travel to Antarctica for their studies and stay there for extended periods since they, too, will be very isolated there and living under very harsh conditions.

NASA has something called the Human Research Program (HRP), which is all about helping astronauts manage their mental health better and keep them sharp and positive. The HRP also focuses on astronauts' physical health. That means it focuses on preventing muscle and bone mass loss (*Human Research Program*, 2019). One of the HRP's most important findings is that spending quality time together outside of "work" can be immensely helpful to the astronaut's mood, mental health, ability to work together, and ability to read one another. This is why they're mandated to engage in various recreational activities together. That can be a hard mandate to live with if you dislike a fellow crewmate, but hopefully, that's not the case, at least not most of the time.

Interestingly enough, food—its quality, taste, and the sentimental connotations that come with it—do a great deal to improve the astronauts mental health than previously thought. This is why NASA and other space agencies create dishes that taste good instead of providing astronauts with the paste-filled tubes that astronauts were originally given when the space program started. It's also why astronauts are encouraged to grow some foods in space, which they do. This ensures they have tasty and healthy ingredients to work with and provides them with an additional recreational activity they can get engaged in together: gardening (Dakkumadugula et al., 2023). Hatch chili peppers, so named because they're grown in hatches, leafy greens, and radishes, are some examples of what astronauts grow together in space. With such ingredients, preparing dishes native to the countries that astronauts are from becomes even easier. In 2021, for instance, astronaut Mark Vande Hei was able to make space tacos, which the entire crew got to enjoy, from the ingredients they had grown themselves. In the process, he broke the record for most foods grown in space.

The Art of Space Living

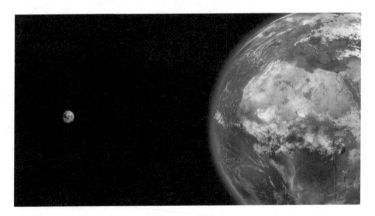

Gardening isn't the only recreational activity ISS crew members engage in. ISS astronauts are known to engage in various recreational activities, especially artistic ones. Art is a good way of processing and dealing with difficult emotions. That's why art therapy is a thing, for instance. This is why astronauts are encouraged to take up artistic practices, be it writing, painting, drawing, or something else. The encouragement that astronauts have received over the years has resulted in the creation of certain artworks that can only ever exist in space. Take the Impossible Object sculpture. A doctor of physics created the concept for this sculpture while on Earth. The sculpture was only on the ISS when a new crew headed up for the AX-1 mission (Jackson, 2022). The sculpture is made of a tapestry of brass tubes and roads. Marbles of self-contained water travel upon these brass tubes and roads.

They can only do so in the microgravity environment of the ISS, though. If the sculpture were to be brought back to the Earth, the water marbles would collapse and drip onto the floor because of the gravity factor on Earth. In the ISS's microgravity environment, these marbles can maintain their shape, stick to the sculpture, and remain there.

Music is another art form that astronauts engage in on the ISS. This is why Adrienne Provenzano wrote a composition dubbed the International Space Station Suite (Provenzano, 2019). The suite comprises seven sections, and a different part or aspect of the ISS has inspired each section. To be clear, Adrienne isn't an astronaut and, therefore, hasn't been to the ISS, but she is a solar system ambassador whom NASA chose. One of Adrienne's sources of inspiration was the cupola installed on the ISS in 2010, allowing astronauts to view the Earth. Plenty of astronauts have given musical performances from this very spot. Candy Coleman played the flute before the cupola during her time on the station. Chris Hadfield played the guitar and sang there, as did Drew Feustel.

One of the most entertaining art-related activities practiced in the ISS is painting, according to the astronauts there. Drawings are relatively easy to do, though no less relaxing or enjoyable. Cosmonaut Alexey Leonov, for instance, used to do orbital sunrise sketches and charcoal portraits of his crewmates. His pictures are hailed as the first to ever be drawn in space (Brown, 2015). However, Nicole Stott did watercolor paintings, which were more challenging than sketches. After all, she couldn't leave her paints out in the open since they would float away, hindering her ability to paint and use them (Stott, 2019). Nicole became the first person to paint watercolors in space. To paint, she would squeeze out a tiny drop of water from a drink bag shaped like a CapriSun bag. The resultant drop of water would hover in the air, and Nicole would move her brush through it. She would then bring the brush toward the piece of paper she was working with without letting it touch it. Instead, she'd drag it just over it, letting the motion sprinkle and spread the color over the paper.

Having said all that, if there's one form of art that all astronauts engage in, it's photography. How could they not, with the beautiful view of the Earth right there, outside the cupola? Astronauts don't just photograph the Earth, though. They also take photos of interesting natural phenomena that they can observe. Astronaut Andreas Morgan, for instance, captured a photo of a thunderstorm on Earth while on the ISS. He did this with a special camera called Davis, which automatically responds to changes in brightness, which came in handy when dealing with a thunderstorm (*International Space Station Gallery*, n.d.). All these photos and more are currently on display in the Space Station Gallery, which is online, though many can be found directly on the astronauts' social media accounts.

Chapter 10:

The Future of the ISS— Missions and New Developments

I think the International Space Station is providing a key bridge from us living on Earth to going somewhere into deep space. —Peggy Whitson

The ISS will only be around for another few years, but the space agencies and countries involved in it aim to make excellent use of it before it's taken out of commission. In the coming years, it will be home to even more exciting, innovative experiments. It will also serve as the base of brand-new missions that will prove to be more ambitious and exhilarating than any before.

Exciting New Missions

One of the key things the ISS will be focusing on in the following years will be sustainability. ISS is already searching for new ways to develop sustainable technologies and achieve a sustainable lifestyle, as you've seen. They will redouble their efforts for this in the years to come. On top of that, the ISS will be focusing on developing new manufacturing techniques and capabilities. These will make it easier to both drive new products to markets more quickly and make space construction a reality, which is something that's necessary for mankind's continued presence on the moon following the construction of the Lunar Gateway, as previously explored.

One of the ISS' main sustainability projects focuses on implementing the sustainable technologies it develops on the planet itself. Dubbed the Design in Space for Life on Earth to Reimagine our Future in Space and done in partnership with the World Design Organization, the project has over 70 participants. In traditional ISS fashion, these participants are from 26 different countries. Things will kickstart on Earth with a team of experts coming up with possible project ideas. Then, spaceflight projects will be initiated so that those ideas can be tested on the ISS' orbiting lab. After that, ideas that prove their worth will be taken down to Earth, where the previous Earth-bound team will start translating them into a reality (*New Project to Apply Sustainable*, 2020).

Another of the ISS's focus points is VR, of all things. The ISS has recently partnered with TIME Studios to explore the VR potential of space. Dubbed *Space Explorers: The ISS Experience*, the project is the largest production ever filmed in space. The idea is to create a VR program that allows participants to experience the first-ever spacewalk firsthand. Participants experience this walk from a first-person POV as if they were living through it personally. This is just step one for the TIME-ISS partnership. The larger goal is to allow the masses to experience the space station and space without having to go there yet.

To that end, professional 3D, 360-degree cameras have been added to the ISS. During filming, astronauts have doubled as both "subjects" and cameramen. So, they had to learn how to shoot the footage that would be used in some parts of the VR experience. They even got to act as creative directors, though they still received direction from the ground, making this the most long-distance filming experience, too. *Space Explorers* is currently out, and it looks to be just beginning. Who knows, perhaps this will be the start of a whole new movie experience, the same way movie theaters were back in the day.

Last but not least, the ISS aims to redefine the global economy, an ambitious goal, to be sure, but well within the realm of possibility. The existing space market is valued in excess of $400 billion and includes industries like space tourism, commercial rockets, ground communication systems, and microsatellites. Over the next two decades, it's expected that this market will keep growing and become worth around $1 trillion. Add to that the planned creation of multiple space stations orbiting the Earth, and you can't help but wonder how much the industry will be worth in the next few years alone. In fact, experts believe that the industry will be worth $1 billion by 2040 (Sheetz, 2022).

Goodbye to the ISS

The ISS has been home to many wondrous discoveries. However, none of these discoveries have proven to be as powerful as how useful microgravity is, at least not in light of how microgravity impacts the development of other experiments. After all, microgravity affects biological factors, like how quickly or slowly bacterial processes unfold in space, a factor that can dramatically change the drug-pharmaceutical markets. This is an effect that needs to be studied more as it indicates that microgravity has a great deal of economic advantages to offer humanity (Stoudemire, 2020). This is especially true in light of how microgravity affects the physical properties of fluids and certain kinds of materials. It can be said it has near-boundless potential for experimentation and innovation. Think about the sheer number of health and technology products that could be created in such an environment.

All this is why a company called Space Tango is working to explore the true potential that microgravity offers and is trying to see what different market segments it can open up. Luckily, their explorations won't be confined to the ISS, at least not once those new space stations are up and running. Eventually, the ISS will have to be de-orbited, and those new space stations will be launched. There are many reasons for this, one being that maintenance operations become more and more costly as time goes on. So, at some point, things will get to where launching new space stations will be more economically feasible than repairing and maintaining the ISS. Another important reason is that the space radiation that the ISS has been subjected to for years is slowly taking its toll on the hull. While the ISS is made with various protective materials to safeguard astronauts' health and well-being from that radiation, eventually, the hull will degrade to a point where those materials and that shielding won't be as effective as they used to be. At that point, keeping astronauts on the ISS simply won't be safe for them anymore.

A final reason why the ISS will be taken down eventually is that constant docking takes its toll on the station, too. Docking a spacecraft on a space station isn't a gentle affair. There's some degree of impact involved (Kluger, 2015). True, there's cushioning in between, and the ISS is made of sturdy stuff, but those impacts take their toll, too, weakening the ISS over time. Maintenance operations do help mitigate this, but again, they can't be done indefinitely. Hence, we will all have to say goodbye to the ISS when the time comes.

Conclusion

The ISS is a wonder in many ways. On the one hand, it's a marvel of human ingenuity, a prime example of the feats that human beings can accomplish when they put their minds to it. On the other, it's a beacon of hope for the world. It's a clear symbol that different nations with different interests and ideologies can work closely together for years on end if they put their mind to it. This is especially true considering the fact that the ISS was the brainchild of the United States and Russia, two former enemies that were on the brink of war mere years before the ISS was officially launched.

In the end, these two countries' choice of collaboration over conflict has been more beneficial for the world and humankind than can ever be put into words. It has led to scientific developments, like medicine to help tackle COVID-19 or new energy systems that use bacteria, of all things, that have and can make a huge difference for the people of Earth. It has led to the development of technologies that might not have been possible or even thought of for years to come, like the new solar arrays that are currently powering most of the station. It has even led to the creation of artworks like actual photographs of the Earth taken from orbit, symphonies written for the ISS, and watercolor paintings done in space that no one would ever have been able to dream up.

The ISS has also been a hub for space exploration. Humankind has learned so much about space from the space station, as well as about how the human body does in space. We've discovered and observed all sorts of interesting things, like how neutron stars are formed and how microgravity impacts the human body. Yet, we still have so much more to discover. The ISS may only be with us for another few years, but there's no question that it will help us make some stupendous discoveries in the meantime. Besides, it's not like the era of discoveries will come to an abrupt end once the ISS is de-orbited. If anything, a new, even faster era of discoveries will begin, where a multitude of experiments are being conducted by a wide variety of space stations. The existence of all these space stations is made possible by the pioneer that is the ISS; the first manmade creation to ever prove, once and for all, that putting a station in orbit and having people live on it wasn't mere fantasy or, rather sci-fi, but instead a huge possibility and ultimately, a reality.

Bonus Chapter:

Strange Facts About the ISS

The ISS has been home to some fascinating discoveries and developments, but it has also seen its fair share of bizarre, funny, and unexpected events and happenings. In light of that, here are some of the strangest, most bizarre facts about the ISS:

- Astronauts on the ISS get to enjoy 16 sunsets a day because the ISS takes just 90 minutes to circle planet Earth (*Astronauts Will Experience*, 2023). Imagine witnessing such a thing rather than waking up early to watch the sunrise!

- There is an upcoming reality TV show planned called *Space Hero* (Jenny, 2022). *Space Hero* has 24 contestants living in a "space village" on Earth, where they get to go through the same training that astronauts go through to head to the ISS. The winner of the reality show will get to go to the ISS and live with the astronauts there for ten days. Their visit will cost $55 million, and this bill will be footed by a company that aims to launch its own private space station in the future since getting a civilian into space like this will help them to set their expectation properly and serve as phenomenal advertising for them, as well as prove to be a demonstration of their capabilities. I wonder who will win?

- Contrary to expectations, the ISS is always leaking air, and these leaks need to be constantly fixed. In 2019, another such leak occurred, and the astronaut who noticed it couldn't find where the leak was located (Neilson, 2020). None of the crew members could, despite searching for it for two months. To solve the matter, cosmonaut Anatoly Ivanishin opened up a packet of tea leaves and let them float in the air. He repeated this process in different parts of the station until the leaves started floating toward a crack in the wall of one specific room. Thus, the leak was identified and fixed with a special tape (yes, tape) that can stick to surfaces where gravity doesn't exist.

- Pizza Hut made an actual delivery to the ISS once. This happened when the cosmonaut Yuri Usachov was craving pizza (*Cosmonauts Eat Pizza*, 2001). NASA agreed to send him pizza if they could find someone to supply it and if Yuri Usachov was willing to wait for two months. Pizza Hut agreed to step up to the challenge, thinking this would be great advertising for them. They had to experiment with various toppings to find one that wouldn't get moldy as the pizza traveled to the ISS. The topping they ultimately went with was salami (pepperoni, sadly, didn't work). Pizza Hut spent $1 million to send a pizza to the ISS, becoming the first restaurant to ever make a delivery to space.

Author Biography

References

About Robonaut. (n.d.). The National Aeronautics and Space Administration. https://www.nasa.gov/robonaut2/what-is-a-robonaut/

Achenbach, J. (2013, September 13). *The International Space Station is one of humanity's great engineering triumphs. But now NASA has to face a difficult question: what is it for?* Washington Post. https://www.washingtonpost.com/sf/national/2013/0 9/14/the-skies-the-limits/

Active thermal control system (ATCS) overview. (2022). Boeing. https://www.nasa.gov/wp-content/uploads/2021/02/473486main_iss_atcs_overv iew.pdf

Astronauts will experience the New Year 16 times while in space. Here's how. (2023, December 31). Hindustan Times. https://www.hindustantimes.com/science/astronauts-will-experience-the-new-year-16-times-while-in-space-heres-how-101704027206148.html#:~:text=Due%20to%20the%2 0space%20station

Azriel, M. (2012, June 14). *Micrometeoroid hit ISS cupola.* Space Safety Magazine. https://www.spacesafetymagazine.com/space-debris/kessler-syndrome/micrometeroid-hit-iss-cupola/

Bartels, M. (2021, November 15). *Space debris forces astronauts on space station to take shelter in return ships.* Space. https://www.space.com/space-debris-astronauts-shelter-november-2021

Benefits for everyone from International Space Station research. (2022, July 24). The National Aeronautics and Space Administration. https://www.nasa.gov/missions/station/iss-research/benefits-for-everyone-from-international-space-station-research/

Blakeslee, R. J., Lang, T. J., Koshak, W. J., Buechler, D., Gatlin, P., Mach, D. M., Stano, G. T., Virts, K. S., Walker, T. D., Cecil, D. J., Ellett, W., Goodman, S. J., Harrison, S., Hawkins, D. L., Heumesser, M., Lin, H., Maskey, M., Schultz, C. J., Stewart, M., & Bateman, M. (2020). Three Years of the Lightning Imaging Sensor Onboard the International Space Station: Expanded Global Coverage and Enhanced Applications. *Journal of Geophysical Research: Atmospheres,* *125*(16). https://doi.org/10.1029/2020jd032918

A brief history of space exploration. (n.d.). The Aerospace Corporation. https://aerospace.org/article/brief-history-space-exploration#:~:text=On%20Oct.

Brown, M. (2015, August 31). *First picture drawn in space to appear in cosmonauts show in London.* The Guardian. https://www.theguardian.com/science/2015/aug/31/first-picture-space-cosmonauts-science-museum-alexei-leonov

CALET: Experiment - International Space Station. (2013, June 10). Japan Aerospace Exploration Agency. https://iss.jaxa.jp/en/kiboexp/ef/calet/

Campbell, A. (2017). Amateur radio on the International Space Station (ARISS). *National Aeronautics and Space Administration.*

Cernan, G. (2020). *Gene Cernan quotes.* GoodReads. https://www.goodreads.com/quotes/6721853-curiosity-is-the-essence-of-our-existence

Chao, T. (2013, May 12). *Space station evolution: 6 amazing orbital outposts.* Space. https://www.space.com/21051-space-station-evolution-history.html

Cosmonauts eat pizza. (2001, May 23). ABC News. https://abcnews.go.com/Technology/story?id=98548&page=1

Cox, R. M., Wolf, J. D., Lieber, C. M., Sourimant, J., Lin, M. J., Babusis, D., DuPont, V., Chan, J., Barrett, K. T., Lye, D., Kalla, R., Chun, K., Mackman, R. L., Ye, C., Cihlar, T., Martinez-Sobrido, L., Greninger, A. L., Bilello, J. P., & Plemper, R. K. (2021). Oral prodrug of remdesivir parent GS-441524 is efficacious against SARS-CoV-2 in ferrets. *Nature Communications, 12*(1). https://doi.org/10.1038/s41467-021-26760-4

Crane, K. W., Corbin, B. A., Lal, B., Buenconsejo, R. S., Piskorz, D., & Weigel, A. L. (2023, May 10). *Report: Market Analysis of a Privately Owned and Operated Space Station (IDA 2017).* New Space Economy. https://newspaceeconomy.ca/2023/05/10/private-space-station/

Cranford, N., & Turner, J. (2021, February 2). *The human body in space.* The National Aeronautics and Space Administration. https://www.nasa.gov/humans-in-space/the-human-body-in-space/

Crespo, D. (2023, May 18). *Cultural cuisines in space: Bringing astronauts together.* Roundup Reads. https://roundupreads.jsc.nasa.gov/roundup/2147

Curtiss, R., Xin, W., Li, Y., Kong, W., Wanda, S.-Y., Gunn, B., & Wang, S. (2010). New Technologies in Using Recombinant Attenuated Salmonella Vaccine Vectors. *Critical Reviews in Immunology, 30*(3), 255–270. https://www.ncbi.nlm.nih.gov/pmc/articles/PMC3970581/

Dakkumadugula, A., Pankaj, L., Alqahtani, A. S., Ullah, R., Ercisli, S., & Murugan, R. (2023). Space nutrition and the biochemical changes caused in Astronauts Health due to space flight: A review. *Food Chemistry: X,* 100875. https://doi.org/10.1016/j.fochx.2023.100875

Datta, A. (2021, June 3). *Damage to Canadarm2 on ISS once again highlights space debris problem.* SpaceNews. https://spacenews.com/op-ed-damage-to-canadarm2-on-iss-once-again-highlights-space-debris-problem/

de Vet, S. J., & Rutgers, R. (2007). From waste to energy: First experimental bacterial fuel cells onboard the international space station. *Microgravity Science and Technology, 19*(5-6), 225–229. https://doi.org/10.1007/bf02919487

Dobrijevic, D., & Jones, A. (2021, August 24). *China's Tiangong space station.* Space. https://www.space.com/tiangong-space-station

Dunham, W. (2023, June 22). Study reveals how immune system of astronauts breaks down. *Reuters.* https://www.reuters.com/technology/space/study-reveals-how-immune-system-astronauts-breaks-down-2023-06-22/

ECOSTRESS. (n.d.). NASA Jet Propulsion Laboratory (JPL). https://www.jpl.nasa.gov/missions/ecosystem-spaceborne-thermal-radiometer-experiment-on-space-station-ecostress

Esen, E. (2020, April 6). *Space-based research to advance technologies for reducing harmful emissions on Earth.* ISS National Laboratory. https://www.issnationallab.org/space-based-research-advance-technologies-reduce-harmful-emissions-earth/

Evans, W. A., & Laufer, D. (n.d.). *Logistics Lessons Learned in NASA Spaceflight.* Interplanetary Supply Chain Management and Logistics Architectures. http://strategic.mit.edu/docs/4_17_NASA-TP-2006-214203.pdf

Everyday life on the ISS. (n.d.). JAXA Human Spaceflight Technology Directorate. https://humans-in-space.jaxa.jp/en/life/health-in-space/life/

Flittner, D. E. (2017, January 12). *About SAGE III on ISS.* SAGE (Stratospheric Aerosol and Gas Experiment). https://sage.nasa.gov/missions/about-sage-iii-on-iss/

Foust, J. (2023a, March 21). *Industry sees missed opportunity in deorbiting ISS.* SpaceNews. https://spacenews.com/industry-sees-missed-opportunity-in-deorbiting-iss/

Foust, J. (2023b, June 8). *From one, many: The race to develop commercial space stations and the markets for them.* SpaceNews. https://spacenews.com/from-one-many-the-race-to-develop-commercial-space-stations-and-the-markets-for-them/

Frank Rubio. (n.d.). The National Aeronautics and Space Administration. https://www.nasa.gov/people/frank-rubio/

Garcia, M. (2018, April 30). *What does it take to keep the station stocked with supplies?* National Aeronautics and Space Administration. https://blogs.nasa.gov/spacestation/2018/04/30/what

-does-it-take-to-keep-the-station-stocked-with-supplies/

Gendreau, K. C., Arzoumanian, Z., & Okajima, T. (2012). The Neutron Star Interior Composition ExploreR (NICER): an Explorer mission of opportunity for soft X-ray timing spectroscopy. *Proceedings of SPIE.* https://doi.org/10.1117/12.926396

Greene, B. (2019, July 11). *Spaceflight studies for a sustainable future.* ISS National Laboratory. https://www.issnationallab.org/spaceflight-studies-for-a-sustainable-future/

Grimm, D., Schulz, H., Krüger, M., Cortés-Sánchez, J. L., Egli, M., Kraus, A., Sahana, J., Corydon, T. J., Hemmersbach, R., Wise, P. M., Infanger, M., & Wehland, M. (2022). The Fight against Cancer by Microgravity: The Multicellular Spheroid as a Metastasis Model. *International Journal of Molecular Sciences, 23*(6), 3073. https://doi.org/10.3390/ijms23063073

Guzman, A. (2022a, July 22). *Understanding the psychological hazards of spaceflight on the space station.* The National Aeronautics and Space Administration. https://www.nasa.gov/missions/station/understanding-the-psychological-hazards-of-spaceflight-on-the-space-station/

Guzman, A. (2022b, July 23). *Improving eye surgery with space technology.* The National Aeronautics and Space Administration. https://www.nasa.gov/missions/station/improving-eye-surgery-with-space-technology/

Guzman, A. (2022c, July 23). *Space station leads to breakthroughs in human health on Earth.* The National Aeronautics and Space Administration.

https://www.nasa.gov/missions/station/space-station-leads-to-breakthroughs-in-human-health-on-earth/

Haupt, A. (2023, March 2). *Your houseplants have some powerful health benefits.* Time. https://time.com/6258638/indoor-plants-health-benefits/

History of astronomy. (n.d.). University of Oregon. https://pages.uoregon.edu/jschombe/ast121/lectures/lec02.html

How the human body changes in space. (n.d.). Baylor College of Medicine. https://www.bcm.edu/academic-centers/space-medicine/translational-research-institute/space-health-resources/how-the-body-changes-in-space#:~:text=In%20space%2C%20astronauts%20may%20face

Howell, E. (2018a, February 8). *International Space Station: Facts, history & tracking.* Space. https://www.space.com/16748-international-space-station.html

Howell, E. (2018b, July 11). *Skylab: First U.S. space station.* Space. https://www.space.com/19607-skylab.html

Howell, E. (2018b, March 26). *Tiangong-1: China's first space station.* Space. https://www.space.com/27320-tiangong-1.html

Howell, E. (2019, September 4). *Scott Kelly: The American astronaut who spent a year in space.* Space. https://www.space.com/32907-scott-kelly-astronaut-biography.html

Howell, E. (2023, August 24). *ISS fires thrusters to avoid oncoming space junk.* Space. https://www.space.com/international-space-station-debris-avoidance-maneuver-august-2023

Huang, B., Li, D.-G., Huang, Y., & Liu, C.-T. (2018). Effects of spaceflight and simulated microgravity on microbial growth and secondary metabolism. *Military Medical Research, 5*(1). https://doi.org/10.1186/s40779-018-0162-9

Human research program. (2019). The National Aeronautics and Space Administration. https://www.nasa.gov/hrp/

International Space Station facts and Figures. (n.d.). The National Aeronautics and Space Administration. https://www.nasa.gov/international-space-station/space-station-facts-and-figures/

International Space Station gallery. (n.d.). The National Aeronautics and Space Administration. https://www.nasa.gov/international-space-station/space-station-gallery/

International Space Station tour. (2012). The National Aeronautics and Space Administration. https://www.nasa.gov/international-space-station/suni-iss-tour/

International Space Station Transition Report. (2022). The National Aeronautics and Space Administration. https://shorturl.at/tLX19

Jackson, B. (2022, November 28). *Impossible object: new art installation only possible in space makes debut on International Space Station.* Glasgow World. https://www.glasgowworld.com/read-this/impossible-objectnew-art-installation-only-possible-in-space-makes-debut-on-international-space-station-3934107

Jenny. (2022, September 29). *Space Hero reality show to build Space Villages around the world.* Globetrender. https://globetrender.com/2022/09/29/space-hero-reality-show-space-villages/

Joel Montalbano, International Space Station program manager. (n.d.). The National Aeronautics and Space Administration. https://www.nasa.gov/people/joel-montalbano-international-space-station-program-manager/

Jupiter: Facts. (2023). The National Aeronautics and Space Administration. https://science.nasa.gov/jupiter/facts/

Keeter, B. (2018). *Spot the station.* National Aeronautics and Space Administration. https://spotthestation.nasa.gov/

Khalid, A., Prusty, P. P., Arshad, I., Gustafson, H. M., Isra Jalaly, Nockels, K., Bentley, B. L., Goel, R., & Elisa Raffaella Ferrè. (2023). Pharmacological and non-pharmacological countermeasures to Space Motion Sickness: a systematic review. *Frontiers in Neural Circuits, 17.* https://doi.org/10.3389/fncir.2023.1150233

Kim, H. J., Olson, D. R., & Laguette, S. (2012). International Space Station Agricultural cameraA (ISSAC) sensor onboard the International Space Station (ISS) and its potential use on the Earth observation. *Dept. Of Earth System Science and Policy, John D. Odegard School of Aerospace Sciences University of North Dakota.*

Kluger, J. (2015, August 27). *How astronauts dock at the space station.* Time. https://time.com/4008222/soyuz-space-station/

Kluger, J. (n.d.). *Meet the twins unlocking the secrets of space.* Time. https://time.com/meet-the-twins-unlocking-the-secrets-of-space/

Kramer, H. J. (2002). *ISS: EMIT (Earth surface mineral dust source investigation).* EoPortal. https://www.eoportal.org/satellite-missions/iss-emit#summary

Kramer, H. J. (2015). *ISS: Imagery from 2015 to 1998*. EoPortal. https://www.eoportal.org/other-space-activities/iss-imagery-from-2015-to-1998

Lea, R. (2023, November 23). *Astronauts celebrate Thanksgiving in space! Here's what they'll eat and what they're thankful for (video)*. Space. https://www.space.com/astronauts-thanksgiving-international-space-station

Lee, S.-J., Lehar, A., Meir, J. U., Koch, C., Morgan, A., Warren, L. E., Rydzik, R., Youngstrom, D. W., Chandok, H., George, J., Gogain, J., Michaud, M., Stoklasek, T. A., Liu, Y., & Germain-Lee, E. L. (2020). Targeting myostatin/activin A protects against skeletal muscle and bone loss during spaceflight. *Proceedings of the National Academy of Sciences, 117*(38), 23942–23951. https://doi.org/10.1073/pnas.2014716117

Lewin, S. (2019, April 11). *Landmark NASA twins study reveals space travel's effects on the human body*. Space. https://www.space.com/nasa-twins-study-kelly-astronauts-results.html

Lewis, R. E. (2023, March 16). *Behavioral health*. The National Aeronautics and Space Administration. https://www.nasa.gov/directorates/esdmd/hhp/behavioral-health/

Love, J. (2023a, July 24). *Science in space: Week of July 17, 2023 - genetic analysis*. The National Aeronautics and Space Administration. https://www.nasa.gov/missions/station/iss-research/science-in-space-week-of-july-17-2023-genetic-analysis/

Love, J. (2023b, November 22). *Inspiring students with ham radio, other educational programs* . The National Aeronautics and Space Administration.

https://www.nasa.gov/missions/station/science-on-station-november-2023/

Lucke, R. L., Corson, M., McGlothlin, N. R., Butcher, S. D., Wood, D. L., Korwan, D. R., Li, R. R., Snyder, W. A., Davis, C. O., & Chen, D. T. (2011). Hyperspectral Imager for the Coastal Ocean: instrument description and first images. *Applied Optics, 50*(11), 1501–1516. https://doi.org/10.1364/AO.50.001501

Malik, T. (2023a, December 3). *Cosmonauts dock Russian Progress cargo ship to ISS on remote control after autopilot glitch.* Space. https://www.space.com/russian-cosmonauts-dock-progress-86-cargo-ship-at-space-station.

Malik, T. (2023b, December 25). *Astronauts in space beam holiday wishes to Earth for Christmas (video).* Space. https://www.space.com/astronanuts-send-christmas-holiday-wishes-2023-video

Medical operations team activities. (n.d.). JAXA Human Spaceflight Technology Directorate. https://humans-in-space.jaxa.jp/en/biz-lab/med-in-space/healthcare/medops/system/

Mir FAQs - facts and history. (2001, February 21). European Space Agency. https://www.esa.int/About_Us/Corporate_news/Mir_FAQs_-_Facts_and_history

Mohon, L. (2020, May 28). *MAXI J1820+070: Black hole outburst caught on video.* The National Aeronautics and Space Administration. https://www.nasa.gov/universe/maxi-j1820070-black-hole-outburst-caught-on-video/

Monitoring the airways. (2017, April 28). European Space Agency. https://www.esa.int/Science_Exploration/Human_and_Robotic_Exploration/Research/Monitoring_the_airways

Moonrise from the ISS. (2012, January 26). European Space Agency. https://www.esa.int/ESA_Multimedia/Images/2012/0 1/Moonrise_from_the_ISS

Mullenweg, M. (n.d.). *Matt Mullenweg quotes.* Academia. https://www.academia.edu/40917330/_Technology_is _best_when_it_brings_people_together_Matt_Mullenw eg_a_Social#:~:text=Mullenweg%2C%20a%20Social-

Munson, O. (2023, May 9). *Is there gravity on the moon? Here's how the satellite's gravity compares to Earth's.* USA Today. https://www.usatoday.com/story/tech/science/2023/ 05/09/gravity-on-the-moon/11727506002/#:~:text=The%20moon%20has %20a%20surface

Murphy, K., & Schauer, K. (2023, October 25). *NASA's first two-way end-to-end laser communications relay system.* The National Aeronautics and Space Administration. https://www.nasa.gov/technology/space-comms/nasas-first-two-way-end-to-end-laser-communications-system/

NASA TV live. (2022, December 7). The National Aeronautics and Space Administration. https://www.nasa.gov/nasatv/

NBL facilities. (2017, May 9). Oceaneering | Connecting What's Needed with What's Next. https://www.oceaneering.com/space-systems/nbl-facilities/

Neilson, S. (2020, October 20). *Space-station crew members just found an elusive air leak by watching tea leaves float in microgravity.* Business Insider. https://www.businessinsider.com/astronauts-cosmonauts-found-space-station-leak-using-tea-leaves-2020-

10#:~:text=The%20International%20Space%20Station %20has

New project to apply sustainable space technologies on Earth. (2020, October 13). Innovation News Network. https://www.innovationnewsnetwork. com/new- project-to-apply-sustainable-space-technologies-on- earth/7416/

NICER (Neutron star interior composition explorer) TELESCOPE. (n.d.). SpaceTV. https://www.spacetv.net/nicer- neutron-star-interior-composition-explorer-telescope/

Nola Taylor Redd. (2017, August 25). Vomit comet: Training flights for astronauts. Space. https://www.space.com/37942- vomit-comet.html

O'Callaghan, J. (2023, May 3). A fiery end? How the ISS will end its life in orbit. BBC. https://www.bbc.com/future/article/20230502-a- fiery-end-how-the-iss-will-end-its-life-in-orbit

Oberhaus, D. (2020, September 9). How Cold War politics shaped the International Space Station. Smithsonian Magazine. https://www.smithsonianmag.com/science- nature/how-cold-war-politics-shaped-international- space-station-180975743/

Patel, S. (2022, March 31). Impact story: Roll-Out solar arrays. The National Aeronautics and Space Administration. https://www.nasa.gov/directorates/stmd/impact- story-roll-out-solar-arrays/

Platt, J. (2020). Communications. Mars Communications Team at NASA's Jet Propulsion Laboratory. https://mars.nasa.gov/mars2020/spacecraft/rover/co mmunications/#:~:text=It%20generally%20takes%20a bout%205

Provenzano, A. (2019, February 12). *Musical windows to the world.* ISS National Laboratory. https://www.issnationallab.org/musical-windows-to-the-world/

Radiation and health. (2023, July 7). World Health Organization. https://www.who.int/news-room/questions-and-answers/item/radiation-and-health

Radiation and life. (n.d.). European Space Agency. https://www.esa.int/Science_Exploration/Human_an d_Robotic_Exploration/Lessons_online/Radiation_an d_life

Russell, J. F., & Klaus, D. M. (2007). Maintenance, reliability and policies for orbital space station life support systems. *Reliability Engineering & System Safety, 92*(6), 808–820. https://doi.org/10.1016/j.ress.2006.04.020

Russia extends cross flight programme with NASA until 2025. (2023, December 28). Reuters. https://www.reuters.com/technology/space/russia-extends-cross-flight-programme-with-nasa-until-2025-roscosmos-2023-12-28/

Schauer, K. (2019, August 19). *Data rate increase on the International Space Station supports future exploration.* The National Aeronautics and Space Administration. https://www.nasa.gov/missions/station/data-rate-increase-on-the-international-space-station-supports-future-exploration/

Science time from space. (2020, October 2). ISS National Laboratory. https://www.issnationallab.org/stem/educational-programs/science-time-from-space/

Sheetz, M. (2022, May 21). *The space industry is on its way to reach $1 trillion in revenue by 2040, Citi says.* CNBC.

https://www.cnbc.com/2022/05/21/space-industry-is-on-its-way-to-1-trillion-in-revenue-by-2040-citi.html

Shirah, B., Bukhari, H., Pandya, S., Ezmeirlly, H. A., Shirah, B., Bukhari, H., Pandya, S., & Ezmeirlly, H. (2023). Benefits of Space Medicine Research for Healthcare on Earth. *Cureus, 15*(5). https://doi.org/10.7759/cureus.39174

Shukla, D. (2023, March 27). *Alzheimer's: Longer telomeres may be linked to better brain health.* Medical News Today. https://www.medicalnewstoday.com/articles/alzheimers-shorter-telomeres-may-be-linked-to-increased-dementia-risk#What-these-results-mean

Sidharth , M. P. (2023, January 16). *International Space Station evaded three collision risks in 2022.* WION. https://www.wionews.com/science/international-space-station-evaded-three-collision-risks-in-2022-553222

Sohn, R. (2021, December 14). *NASA scientists consider the health risks of space travel.* Space. https://www.space.com/nasa-scientists-health-risks-spaceflight-space-exploration

Space startup Vast partners with SpaceX to launch commercial space station. (2023, May 11). The Economic Times. https://economictimes.indiatimes.com/tech/technology/space-startup-vast-partners-with-spacex-to-launch-commercial-space-station/articleshow/100163145.cms

Space station ambassador program. (n.d.). ISS National Laboratory. Retrieved December 29, 2023, from https://www.issnationallab.org/stem/ambassadors/

Space stations. (2022, March 22). Science Fiction Encyclopedia. https://sf-encyclopedia.com/entry/space_stations

Space telescopes reveal previously unknown brilliant X-ray explosion in Milky Way Galaxy. (2010, October 25). Astronomy

Magazine. https://www.astronomy.com/science/space-telescopes-reveal-previously-unknown-brilliant-x-ray-explosion-in-milky-way-galaxy/

SpaceRef Editor. (2002, February 25). *Effects of EVA and long-term exposure to microgravity on pulmonary function (PuFF): Results.* SpaceRef. https://spaceref.com/status-report/effects-of-eva-and-long-term-exposure-to-microgravity-on-pulmonary-function-puff-results/

Stefanov, W. L. (n.d.). *Space station agricultural camera observes flooding in North Dakota.* European Space Agency. https://www.esa.int/Science_Exploration/Human_an d_Robotic_Exploration/International_Space_Station_ Benefits_for_Humanity/Space_Station_Agricultural_C amera_Observes_Flooding_in_North_Dakota

Stenzel, C. (2016). Deployment of precise and robust sensors on board ISS—for scientific experiments and for operation of the station. *Analytical and Bioanalytical Chemistry, 408*(24), 6517–6536. https://doi.org/10.1007/s00216-016-9789-0

Stott, N. (2019, March 23). *What it's like to paint in space—according to a NASA astronaut.* Quartz. https://qz.com/1578231/what-its-like-to-paint-in-space-according-to-a-nasa-astronaut

Stoudemire, J. (2020, October 21). *Open orbit—expanding the definition of.* ISS National Laboratories. https://www.issnationallab.org/open-orbit-expanding-the-definition-of-global/

Suarez, D. (n.d.). *Daniel Suarez quotes.* GoodReads. https://www.goodreads.com/quotes/tag/outer-space

Sunita L. Williams. (n.d.). The National Aeronautics and Space Administration. https://www.nasa.gov/people/sunita-l-williams/

The Guardian Staff. (2023, July 26). NASA briefly loses contact with ISS after power outage and relies on backup systems for first time. *The Guardian*. https://www.theguardian.com/science/2023/jul/26/iss-power-outage-nasa-contact-international-space-station-backup-systems#:~:text=The%20outage%20meant%20mission%20control

Toothman, J. (1970, January 1). *How space farming works*. HowStuffWorks. https://science.howstuffworks.com/space-farming.htm#:~:text=The%20main%20benefit%20and%20purpose

Uri, J. (2020, December 23). *Space station 20th: Celebrating the holidays in space*. The National Aeronautics and Space Administration. https://www.nasa.gov/history/space-station-20th-celebrating-the-holidays-in-space/

Uri, J. (2021, April 19). *50 years ago: Launch of Salyut, the world's first space station*. The National Aeronautics and Space Administration. https://www.nasa.gov/missions/station/50-years-ago-launch-of-salyut-the-worlds-first-space-station/

Wagenen, V. (2020, January 13). *The ISS engineering feat: Power and cooling*. ISS National Laboratory. https://www.issnationallab.org/the-iss-engineering-feat-power-and-cooling/

Wall, M. (2023, April 27). *Russia agrees to stay aboard International Space Station through 2028*. Space. https://www.space.com/russia-stay-international-space-station-partner-2028

Werner, D. (2023, May 23). *Orbital Sidekick acquires first light imagery*. SpaceNews. https://spacenews.com/orbital-sidekick-acquires-first-light-imagery/

What is Galileo? (n.d.). European Space Agency. https://www.esa.int/Applications/Navigation/Galileo/What_is_Galileo

Whitson, P. (n.d.). *Peggy Whitson quotes*. Brainy Quotes. https://www.brainyquote.com/quotes/peggy_whitson_912890?src=t_international_space_stationhttps://www.brainyquote.com/quotes/peggy_whitson_912890?src=t_international_space_station

Wilke, D., Padeken, D., Weber, Th., & Gerzer, R. (1999). Telemedicine for the International Space Station. *Acta Astronautica, 44*(7-12), 579–581. https://doi.org/10.1016/s0094-5765(99)00065-x

Yang, J.-W., Zhang, M.-X., Ai, J.-L., Wang, F., Kan, G.-H., Wu, B., & Zhu, S.-Q. (2022). Spaceflight-associated neuro-ocular syndrome: a review of potential pathogenesis and intervention. *International Journal of Ophthalmology, 15*(2), 336–341. https://doi.org/10.18240/ijo.2022.02.21

Young, J. (n.d.). *John Young quotes*. Sea Sky. http://www.seasky.org/quotes/space-quotes-space-travel.html

Image References

12019 (2013, March 12). *Japanese space station* [Image]. Pixabay. https://pixabay.com/photos/japanese-space-station-space-92202/

51658 (2016, November 1). *Universe earth radiation* [Image].

Pixabay. https://pixabay.com/photos/universe-earth-planet-space-cosmos-1784292/

Kamil Grygo. (2020, September 21). *Aurora borealis* [Image]. Pixabay. https://pixabay.com/photos/iceland-aurora-borealis-silhouettes-5586225/

Konstantin Koloskov. (2017, February 1). *Analysis* [Image]. Pixabay. https://pixabay.com/photos/analysis-biochemistry-biologist-2030261/

NASA Imagery. (2010, December 5). *Astronaut international space station* [Image]. Pixabay. https://pixabay.com/photos/astronaut-international-space-station-976/

NASA Imagery. (2010, December 5). *International space station iss* [Image]. Pixabay. https://pixabay.com/photos/international-space-station-iss-nasa-988/

NASA Imagery. (2010, December 5). *Spacewalk* [Image]. Pixabay. https://pixabay.com/photos/space-walk-astronaut-nasa-aerospace-991/

P.T. Norbert. (2017, August 1). *Star trails* [Image]. Pixabay. https://pixabay.com/photos/star-trails-astronomy-wallpaper-2234343/

SpaceX Imagery. (2015, March 26). *Mission control* [Image]. Pixabay. https://pixabay.com/photos/space-center-spacex-control-center-693251/

SpaceX Imagery. (2015, March 26). *Satellite orbit* [Image]. Pixabay. https://pixabay.com/photos/satellite-orbit-spacex-aeronautics-693203/

SpaceX Imagery. (2015, March 26). *Spacecraft spacex* [Image]. Pixabay. https://pixabay.com/photos/spacecraft-spacex-spaceship-693217/

The Happy Camper. (2022, September 22). *Earth.* [Image.] Pixabay. https://pixabay.com/photos/earth-world-globe-3d-earth-render-7471157/

Trevor Rock. (2023, March 15). *International space station sunset* [Image]. Pixabay. https://pixabay.com/photos/international-space-station-sunset-7850605/

WikiImages. (2011, December 14). *Astronaut in space suit* [Image]. Pixabay. https://pixabay.com/photos/astronaut-space-suit-space-universe-11080/

WikiImages. (2013, January 4). *Astronaut international space station* [Image]. Pixabay. https://pixabay.com/photos/astronaut-international-space-station-67639/

WikiImages. (2012, December 21). *International space station* [Image]. Pixabay. https://pixabay.com/photos/international-space-station-63128/

WikiImages. (2013, January 4). *International space station eat* [Image]. Pixabay. https://pixabay.com/photos/international-space-station-eat-67774/

WikiImages. (2012, January 9). *Space shuttle to me.* [Image.] Pixabay.

WikiImages. (2011, December 14). *Space station* [Image]. Pixabay. https://pixabay.com/photos/eat-space-station-11114/

Made in the USA
Columbia, SC
18 May 2024

35857433R00080